Introductory Concepts of Biology

George C. Becker
Metropolitan State College
Denver, Colorado

The Macmillan Company New York
Collier-Macmillan Limited London

Copyright © 1972, George C. Becker
Printed in the United States of America

All rights reserved. No part of this book may be reproduced or transmitted in any form or by any means, electronic or mechanical, including photocopying, recording, or any information storage and retrieval system, without permission in writing from the Publisher.

THE MACMILLAN COMPANY
866 Third Avenue,
New York, New York 10022
Collier-Macmillan Canada, Ltd.,
Toronto, Ontario

Library of Congress catalog card number: 72-151170

First Printing

to my parents

WHO HAD THE FORESIGHT
TO GUIDE THEIR FOUR SONS
INTO HIGHER EDUCATION

Preface

This text is intended for the student with little or no background in biology or chemistry. It has, therefore, encountered the problem inherent in the writing of a short introductory text—finding an acceptable compromise between technical depth and ease of understanding for the student. The majority of the biology texts available are either paperbacks dealing with specific aspects of the field or comprehensive reference texts. Both of these types are valuable and find their place in the classroom. There is, however, a need for short texts which present the basic concepts of modern biology with technical accuracy but with a scope which can be covered in a one-term course. Whatever the difficulties of presenting a highly involved and technical subject in limited space, the effort is worthwhile if it results in the communication to the student of a meaningful understanding of the principles of biology upon which he may base further studies of biology or related fields.

This text was written with this aim in mind. The information presented is in a step by step sequence so that the beginner can progress from one consideration to another without being expected to draw on background. It has been the experience of the author that such an approach is most effective in communicating biological concepts to the student who has no previous training in biology or chemistry. The information and explanations are not particularly detailed, but references for more extensive information are given. The student desiring more information may pursue it to a degree in lectures and to a greater degree in outside references. This book is not intended as a scholarly treatment of these subjects, but as an introductory guide for their study.

The purpose here is to present introductory concepts of biology that have broad applications to other fields, to everyday living, or to advanced biology. More and more departments are orienting their beginning courses to such concepts, and following them with courses introducing the organismal approach and taxonomy. It is recommended, however, that all common applications of basic concepts be included in the course when this text is used, especially for nonmajor students.

I wish to acknowledge all my colleagues who have assisted me during the preparation of this manuscript, but especially the following: Dr. Carol Dempster, for her excellent work as a technical reviewer; Dr. A. E. Vatter, University of Colorado, Webb Waring Institute, and Dr. Charles E. Flickenger, University of Colorado, Department of Molecular, Cellular, and Developmental Biology, for their excellent electron micrographs used herein; the personnel of the Denver Federal Center, especially Ottey M. Bishop of the U. S. Bureau of Mines and Robert D. Sullivan and George Hinton of the U. S. Forest Service, for their help; the Colorado Game, Fish, and Parks Division for numerous fine photos; and to Dr. Jean A. Bowles, Dr. John Martin, Dr. Jack Cummins, Dr. Fred Dewey, Dr. Jeff Hurlbut, Dr. John Krenetsky, Dr. Daryl Petersen, Dr. Robert Cohen, and Mrs. Carol S. Steele, of the Metropolitan State College staff, and Dr. Rollie Schaefer, of the New Mexico Institute of Mining and Technology, for their general encouragement and assistance. I also wish to acknowledge my wife's assistance and patience during the preparation of this manuscript, and the many helpful comments of my students during the years in which this text was being developed.

G. C. B.

Contents

Brief Contents

Part I
The Origin of Life and the Cell

1   The origin of life and the evolution of the cell   3
2   Cell structure   29
3   Chemical composition of cytoplasm   49

Part II
Cell Function

4   The role of DNA in cell function   67
5   Protein synthesis   78
6   Photosynthesis   92
7   Respiration   99
8   The movement of substances into and out of cells   110

Part III
Genetics

9   Chromosomes   121
10   Cell division and the life cycle   130
11   An introduction to genetics   148

Part IV
Evolution

12 The role of the organism in evolution 169
13 The role of the environment in evolution 177
14 Patterns in evolution 185

Part V
Energy Flow, Cybernetics, and Population Dynamics

15 Thermodynamics 199
16 Cybernetics and systems control 208
17 Animal population dynamics 215
18 Human population dynamics 226

Part VI
Human Ecology and Conservation

19 Man and air 239
20 Man and water 255
21 Man and energy 270
22 Man and shelter 283

Part VII
On Life, Biology, and Science

23 On life, biology, and science 295

Glossary 307 / **References** 319 / **Index** 322

Detailed Contents

Part I
The Origin of Life and the Cell

1   The Origin of Life and the Evolution of the Cell   3
Level of organization I: The atom   4
Level of organization II: The simple molecule   7
Level of organization III: The complex molecule   14
Level of organization IV: The macromolecule   20
Level of organization V: The cell   21

2   Cell Structure   29
Intracellular structure   31
Extracellular structure   45
The average cell   46

3   Chemical Composition of Cytoplasm   49
Inorganic chemicals in cytoplasm   49
Organic constituents of cytoplasm   51
Elemental composition of cytoplasm   46

Part II
Cell Function

4   The Role of DNA in Cell Function   67
Relative stability of DNA   68
Replication of DNA   68
The coding system of DNA   69

5   **Protein Synthesis**   78
    The mechanism of protein synthesis   78
    The importance of proteins to the cell   81
    Control of protein synthesis   87

6   **Photosynthesis**   92
    Photosynthetic requirements   93
    The process of photosynthesis   94

7   **Respiration**   99
    Glycolysis   100
    Krebs' citric acid cycle   101
    Electron transfer systems   103
    ATP production during glycolysis and respiration   104
    Fermentation   104
    Anaerobic production of ATP by muscles   105
    Production of ATP from general food compounds   106

8   **The Movement of Substances Into and Out of Cells**   110
    Principles of diffusion   110
    Diffusion through a membrane   112
    Movement through a cell membrane   114

Part III
Genetics

9   **Chromosomes**   121
    Chromosomal replication   122
    Homologous chromosomes   123
    Chromosomal mutations   123
    Point mutations   126

10 **Cell Division and the Life Cycle**   130
    Mitotic cell division   130
    Meiotic cell division   135
    The role of cell division in life cycles   145

11 **An Introduction to Genetics**   148
    Mendelian genetics   148
    Lack of dominance   155
    Gene linkage and genetic crossover   155

Practice exercises in genetics 159
Answers to practice exercises 164

Part IV
Evolution

12 **The Role of the Organism in Evolution** 169
    Reproductive potential of populations 170
    Variability of organisms 171
    Variability in organisms and populations 174

13 **The Role of the Environment in Evolution** 177
    Natural selection in stable environments 178
    Effects of environmental change 179
    Isolation-speciation 180
    Changing gene frequencies in evolution 182

14 **Patterns in evolution** 185
    Competitive exclusion 185
    Adaptive radiation 186
    Divergent evolution 187
    Convergent evolution 193
    The time pattern of evolution 194

Part V
Energy Flow, Cybernetics, and Population Dynamics

15 **Thermodynamics** 199
    The first law of thermodynamics 200
    The second law of thermodynamics 204
    Entropy 204
    Open and isolated energy systems 206

16 **Cybernetics and Systems Control** 208
    Systems control in a mechanical system 208
    Systems control in biological systems 212

17 **Animal Population Dynamics** 215
   Reproductive potential 215
   Energy requirements 217
   Energy flow and population size 220
   Systems control factors in a biological community 223

18 **Human Population Dynamics** 226
   Pressures of reproductive potential 226
   Energy requirements of humans 227
   Environmental changes caused by energy use 229
   Systems control factors in human populations 232

Part VI
Human Ecology and Conservation

19 **Man and Air** 239
   Introduction 239
   Air 240
   Sources and effects of air contamination 242
   The future 251

20 **Man and Water** 255
   Biochemical oxygen demand 256
   Eutrophication 257
   Chemical contamination 258
   Persistent contaminants 259
   Biological concentration 260
   Water treatment 261
   The future 267

21 **Man and Energy** 270
   Food 270
   Fuels 276
   The future 279

22 **Man and Shelter** 283
   Clothing 283
   Housing 286
   The future 289

Part VII
On Life, Biology, and Science

23 On Life, Biology, and Science 295
   What is life? 295
   Is a virus living? 297
   What is science? 302

Glossary 307 / **References** 319 / **Index** 322

Part I

The Origin of Life and the Cell

Spiral nebula in Ursa Major, Messiar. (*Photograph from Mount Wilson and Palomar Observatories.*)

The Origin of Life and the Evolution of the Cell 1

One way to begin a study of biology is to retrace the paths by which life probably originated. This approach focuses on the components of living things and how these components fit together to form the structures and perform the functions now associated with "life." Since this is precisely the focus of present-day biological research, an evolutionary approach to the study of biology naturally leads to current biological concepts. It is important to remember that any such description of the development of living things from nonliving matter is merely a plausible *theory*. This theory as to how the living systems now found on earth might have arisen is based on current knowledge of how matter behaves and conjectures as to the conditions on earth during the last 5 billion years.

The physical and chemical processes and events that led to the development of cells from nonliving chemicals are sometimes referred to as chemical evolution. The components of these chemicals are believed to have been present in the solar system early in the history of the earth. As conditions on earth changed, so did the organization of these chemical components. The original components were quite simple. As the earth aged, however, these simple components interacted with one another and formed more complex structures. This chemical development has been divided arbitrarily into levels of organization in this text. The lowest level of organization refers to the simplest chemical components and the higher levels to the more complex arrays of these components. The use of levels of organization is only to aid in classifying the kinds of structures that developed and the general conditions that were necessary for their development. It does not imply that all components rearranged at the same time or that the earth underwent distinct phases of development.

Level of Organization I: The Atom

The primitive earth is believed to have originally existed as a hot mass of atomic gases. The earth today does not resemble that atomic mass, but all matter on it is *composed* of atoms in various combinations. The combinations of atoms found on the earth today would not have been stable at the elevated temperatures of the primitive earth. The atom, however, is exceedingly stable and could exist at those temperatures. Under the conditions on earth around the time of its formation, the atom was the most complex chemical structure.

The Structure of Atoms

All atoms are composed basically of three "subatomic" particles: protons, electrons, and neutrons. Atoms that differ are composed of different numbers of subatomic particles, but there are general rules as to the number and arrangement of these particles within an atom that hold for all the different atoms. These rules result from the specific properties of these subatomic particles. Protons (p^+) possess a unit positive electric charge, electrons (e^-) possess a unit negative electric charge, and neutrons (n) are electronically neutral. Furthermore, neutrons and protons are very similar in weight. Electrons, however, are much smaller and weigh only 1/1845 as much as either a proton or neutron.

All atoms have as many positively charged protons as they have negatively charged electrons. An atom, then, is electronically neutral. The number of protons or electrons, or electron-proton pairs of an atom, is the atomic number of that atom. Naturally occurring atoms have atomic numbers from 1 to 92. The number of neutrons in an atom can vary, but it is generally similar to, or greater than, the atomic number. In general, the higher the atomic number, the greater is the margin by which the number of neutrons exceeds the atomic number. For example, the simplest atom is the atom of hydrogen, atomic number 1 (Fig. 1-1). It is composed of only one proton, one electron, and no neutrons. An atom of helium, atomic number 2, has two electrons, two protons, and two neutrons (Fig. 1-2). At the other end of the atomic spectrum, however, is uranium, atomic number 92. An atom of uranium is composed of 92 electrons, 92 protons, and 146 neutrons.

Protons and neutrons are approximately equal in weight and are much, much heavier than electrons. If the weight of either a proton

p represents a proton

n represents a neutron

e represents an electron

represents an electron shell

represents nucleus of atom

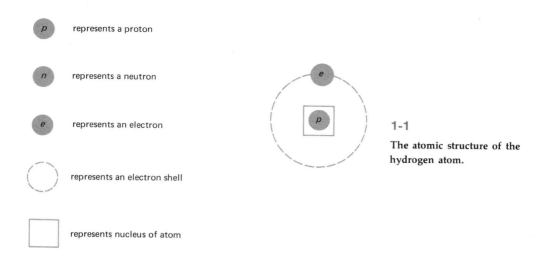

1-1

The atomic structure of the hydrogen atom.

or neutron is taken as 1, a relative scale of atomic weights can be determined. The weight of an atom on this scale is roughly equal to the sum of the protons and neutrons. The atomic weight of hydrogen is 1, helium is 4, and uranium is 238.

For simplicity each atom has a chemical symbol to designate it. The symbol for hydrogen is H, for helium is He, and for uranium is U. For further identification, the atomic weight is designated as a superscript and the atomic number as a subscript. The complete chemical designations for these three atoms, then, is ^1_1H, ^4_2He, and $^{238}_{92}U$.

The atom is so stable because the arrangement of the charged subatomic particles which compose it is such that the attractive forces ($e^- - p^+$) are maximized, and the repulsive forces ($p^+ - p^+$ and $e^- - e^-$) are minimized. In all atoms the more massive protons and neutrons are located in a central core, the nucleus. The smaller, lighter electrons travel at high velocities in the space around the nucleus. The volume that the electrons occupy is much, much larger

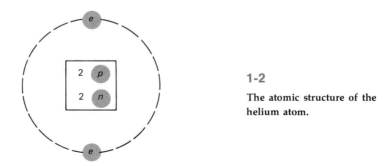

1-2

The atomic structure of the helium atom.

than that of the nucleus. Clearly, the forces on an electron will depend on its distance from the distant but attracting nucleus and its distance from the other repelling electrons which are close to it. An electron, then, will possess a certain level of energy, depending upon the position of the region it occupies relative to the nucleus and other electrons. It is convenient to group regions in which the occupying electrons have very similar energy levels into "shells." As will be discussed shortly, this designation is useful in predicting the chemical reactivity of atoms. Basically, each shell represents electrons within certain regions around the nucleus. The farther a region is from the nucleus, the higher is the energy level for an electron in that region. Therefore, shells of higher energy represent electrons farther from the nucleus.

Each shell can only contain a certain number of electrons. The higher-energy shells can contain more electrons than the lower-energy shells which are closer to the nucleus. The first shell (lowest energy) can only hold 2 electrons. The next two shells can each hold 8 electrons. The successively higher energy shells can hold 18, 18, and 32 electrons, respectively. Shells that can hold 18 electrons can be further divided into two subshells with capacities for 8 and 10 electrons. Those shells that can contain 32 electrons consist of subshells that can hold 8, 10, and 14 electrons. The subshell with 8 electrons is always the outermost shell. The outer electron shell, then, of all atoms except H and He can hold 8 electrons. The outer electron shell for both H and He can only hold 2 electrons.

Because the energy levels of electrons can be grouped in shells, an atom is often depicted as a solar system with the nucleus as the central body and the electrons in rings around it. Each ring represents an electron shell (Figs. 1-1 and 1-2). It is important to note that this diagram is only a conceptual convenience to indicate the relative energy levels of the electrons. It does not mean that the electrons are "orbiting" around the nucleus in fixed "orbits."

It is this electronic configuration of the protons and electrons within an atom that dictates how that atom interacts with other atoms which are either identical or different. It is the myriads of these microscopic interatomic interactions which give rise to the observable macroscopic properties of the substance composed of these atoms. A substance composed of only one kind of atom is an element. An atom, then, is the smallest unit of an element. Any further subdivision would only lead to the three subatomic particles which are identical for all elements. The electronic configuration of an atom with a certain atomic number differs from that of any other atom with a different atomic number since these two atoms contain different numbers of charged subatomic particles. Therefore, no two atoms of different atomic numbers will interact in an identical way with all other atoms. There are 92 naturally occurring atoms with different atomic numbers; there are, therefore, 92 different naturally occurring elements.

Atoms, however, do exist which have the same atomic number but different atomic weights. The heavier atoms have more neutrons than the lighter ones of the same atomic number. Since they only differ in the number of electronically neutral particles, these atoms still behave identically chemically, that is, in the way they interact with other atoms. Atoms with the same atomic number but different atomic weights are called isotopes.

The early earth, then, was composed only of elements. This system was very hot so that most of these elements were in the gaseous state. When heat energy is added to a substance, it causes the atoms of that substance to move faster. The energy of a moving object is called kinetic energy. The faster an object is moving, the higher is its kinetic energy. Therefore, when a substance is heated, the atoms of the substance gain kinetic energy. If atoms are moving fast enough, any attractive interactions they might have toward one another are not strong enough to restrain their motion away from one another. This was apparently the state of the early earth. It was so hot that atoms were moving too rapidly for any attractive interactions between them to cause stable associations. When atoms in such frantic motion collided, they merely ricocheted off one another. The atom, however, because it is composed of such strongly attracting particles, was stable under those conditions. The interactions between the subatomic particles were strong enough to keep them together in spite of the high velocities of these particles.

This picture of the early earth is quite different from the earth as it exists today. Yet it is from these atoms that the world of today, both the living and nonliving forms, arose.

Level of Organization II: The Simple Molecule

This mass of hot gases of the early earth was in space. Space, however, is very cold. This earth system then lost heat to its surroundings. As the system cooled, the atomic movement slowed down. Once the atoms were not in such frantic motion, other forces which were acting on them became important. These forces had not been able to restrain the hotter atoms, but as these atoms lost kinetic energy they were more easily restrained.

One restraining force which became important was gravity. The atoms of the earth mass gradually formed levels according to their atomic weights. The heavier atoms such as iron, ^{56}Fe, and manganese, ^{55}Mn, were pulled toward the center of this mass, leaving the lighter elements, hydrogen, ^{1}H, nitrogen, ^{14}N, carbon, ^{12}C, and oxygen, ^{16}O, in a surrounding outer layer. Intermediate layers were composed of substances with intermediate atomic weights, such as silicon, ^{28}Si, and aluminum, ^{27}Al. The atoms present on the surface of the earth, hydrogen, carbon, nitrogen, and oxygen, are those which were important in chemical evolution, since life apparently developed on the earth's surface.

The other forces acting on these atoms, which became more and more important as the atoms lost kinetic energy, were the interatomic interactions. These interactions brought about two important changes, the formation of molecules and the change of physical state.

The Formation of Molecules

Once the atoms had slowed down, they began to interact with one another when they collided rather than bounce off one another. In some collisions the attractive interactions the atoms experienced were strong enough to hold the atoms together in a molecule. When atoms interact to form such stable associations, this is called a chemical reaction. The force holding atoms together in a molecule is called a chemical bond.

A molecule, then, is a stable composite of atoms held together by a new electronic configuration for this composite system. This composite, therefore, will interact with other atoms or molecules differently than either of its component atoms. The substance which it composes will have unique properties. The molecule is the smallest unit of that substance, as any further subdivision would only break it down into the component atoms. If the molecules of a substance are made up of different atoms, that substance is called a compound. Most molecular substances are compounds, but some atoms such as hydrogen, oxygen, and nitrogen combine with themselves to form the molecular elements H_2, O_2, and N_2, respectively. In chemical shorthand, molecules are represented by a chemical formula which contains the symbols for all the atoms present in the molecule. In these chemical formulas, the subscript of each atomic symbol designates the number of atoms of that type within the molecule. For instance, H_2O, the chemical formula for water, indicates that each molecule of water is composed of two hydrogen atoms and one oxygen atom.

Molecules will form spontaneously if the composite structure of the two or more atoms is at a lower energy state than the individual atoms. If molecules do form spontaneously, the extra energy which the atoms possess relative to the molecule will be given off as either heat or light. If atoms have cooled to a point that a set of individual atoms is at a lower level of energy than the molecule they could form, then they will not form that molecule spontaneously. However, if energy is added to these atoms (either as heat or light) so that the set of atoms is elevated to a level of energy that is higher than the molecule, they can form the molecule.

On the early earth the atoms were so energetic that molecules formed spontaneously. Later on, however, as the earth continued to cool, the atoms (or molecules) did not always possess enough energy to form different structures spontaneously when they came into contact. Molecules did form between these less energetic species, however, when an energy source was available. Molecules formed in this way are "storing" that energy in their chemical bonds. This energy is given off when the molecule is broken down into its original atoms or smaller groups of atoms. This storage of energy in chemical bonds was, and is, very important to the development and maintenance of life.

When atoms interact to form a molecule, the electronic configuration around each atom in the molecule differs somewhat from that of the unbound atoms. Generally, the difference is slight and only the positions of the electrons in the outer electron shell are changed. Molecular formation, therefore, does not alter the basic structure of atoms, only these outer electrons are involved in the chemical bond. The outer shell (or subshell) of most atoms can only hold eight electrons. Depending on its atomic number, an atom has any number from one to eight electrons in its outer shell. Apparently atoms are more stable when their outer electron shells are filled with electrons. When atoms bond together to form molecules, their outer electrons rearrange so that each atom in the molecule ends up with an electron-filled outer shell. There are two major ways in which atoms bond together to attain this stable form. They form ionic bonds, and they form covalent bonds.

Ionic Bonds An ionic bond is formed when one atom "gives away" electrons to another atom. By this process both atoms end up with filled outer shells. An atom with only a single electron

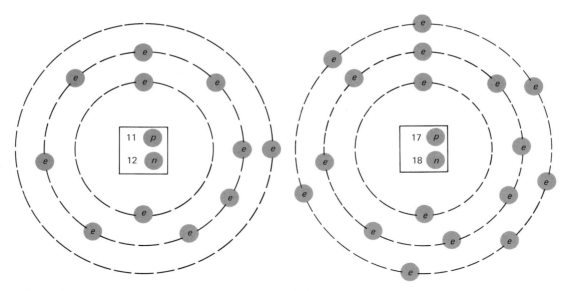

1-3 The atomic structure of the sodium atom.

1-4 The atomic structure of the chlorine atom.

in its outer electron shell readily gives up that electron to an atom that is lacking just one electron in its outer shell. For example, a sodium atom, $_{11}$Na, has a single electron in its third shell and a filled second shell (Fig. 1-3); a chlorine atom, $_{17}$Cl, has seven electrons in its third shell (Fig. 1-4). The chlorine atom lacks only one electron from a complete set of eight electrons in its outer shell. If the sodium atom gives an electron to the chlorine atom, a very

1-5 The exchange of an electron from sodium to chlorine in the formation of an ionic bond.

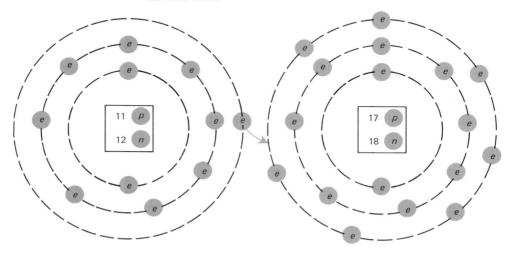

[10]   The Origin of Life and the Cell

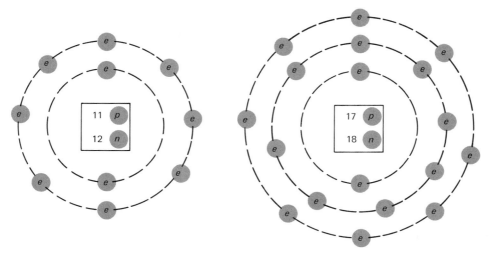

1-6 The electrostatic basis of the ionic bond. The sodium ion at left has a net charge of +1 and the chloride ion at right has a net charge of −1; these ions are attracted to each other by the unlike charges. This is their only bonding force.

stable configuration is attained (Fig. 1-5). By this transfer the sodium becomes positively charged and the chlorine negatively charged. These charged atoms (written Na^+, Cl^-) are called ions. Both ions have filled outer shells (Fig. 1-6).

The mutual attraction between these two oppositely charged ions forms a strong ionic bond. Sodium chloride (NaCl, common table salt) is a very stable compound.

Covalent Bonding When two atoms form a covalent bond, some of the outer electrons rearrange to be "shared" by both atoms. Essentially what this means is that each "shared" electron feels electronic attraction from the nuclei of both bonding atoms. Every "shared" electron, then, helps to fill the outer electron shell of both atoms. In this way the outer electron shells of both atoms are filled without either atom "giving up" an electron as it would to form an ionic bond. Water, H_2O, is a compound with covalent bonds between hydrogen and oxygen atoms. The two hydrogen atoms each only have a single electron (Fig. 1-1). A filled outer shell for hydrogen has only two electrons, so each hydrogen atom needs one electron to complete its outer shell. The oxygen atom has only six electrons in its outer shell (Fig. 1-7). This shell needs two more electrons to be completely filled. If two hydrogen atoms share their electrons with an oxygen atom which in turn shares one of its

The Origin of Life and the Evolution of the Cell **[11]**

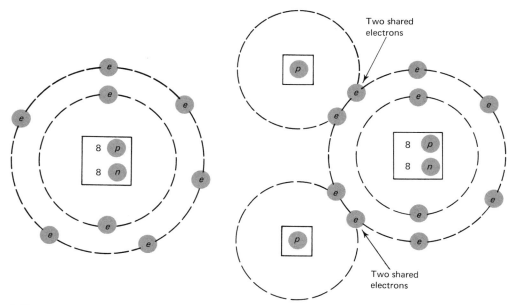

1-7 The atomic structure of the oxygen atom.

1-8 The molecular structure of H_2O or water. The bonding of the hydrogen and oxygen atoms of the H_2O molecule involves sharing electrons between the atoms and is called covalent bonding.

electrons with each hydrogen atom, both the hydrogen atoms and the oxygen atom will have filled outer shells (Fig. 1-8). Covalent bonds are quite strong. Water is a very stable molecule; it forms explosively from O_2 and H_2.

Early Formation of Compounds When the earth system cooled to a point where atoms could interact with one another, both ionic and covalent molecules formed. Theoretically, the atoms that were most reactive and most prevalent in this earth system would get incorporated into the most molecules. The atoms which were the most active and available at the surface of the earth were hydrogen, oxygen, carbon, and nitrogen. The compounds which they undoubtedly formed were methane (CH_4), water (H_2O), ammonia (NH_3), and carbon dioxide (CO_2) (Fig. 1-9).

Physical State and Evolution As the gaseous mass continued to cool, other changes on earth besides the formation of molecules gradually occurred. Just as atoms in a gas behave independently of one another until they are so sluggish that interactions between them can compete with their motions, molecules in a gas behave

[12] The Origin of Life and the Cell

independently until their speed decreases to a point where their attractions toward one another can restrain their motion. In general, intermolecular forces are much weaker than the interatomic forces which led to the formation of molecules. Therefore, the interaction of molecules is only important as much lower temperatures than those at which molecular formation occurred. The interaction of molecules leads to changes of state.

Matter can exist in three physical states: gas, liquid, and solid. The difference between these three states lies in the amount of kinetic energy that the molecules of each state have.

Molecules which have so much kinetic energy that they can overcome any mutual attraction for one another and move independently of one another are in the gaseous state. They are freely separate of one another and can move into any space open to them.

If the molecules of a gas are cooled enough, their rate of travel will slow to a point where intermolecular attractions successfully compete with and hinder independent molecular motion. These intermolecular attractions cause the molecules to get much closer together. This is the liquid state and occupies much less volume than the gaseous state. At temperatures where a substance is liquid, the molecules of the substance cannot break away from one another but they can still move among themselves. Since the molecules can still move among themselves, the liquid state does not have a definite shape. This state appears "runny" and takes the shape of any container.

The solid state of matter results when the compound is cooled

Hydrogen + hydrogen ⟶ H—H   or hydrogen

Carbon + 4 hydrogen ⟶ H—C(H)(H)—H   or methane

2 Hydrogen + oxygen ⟶ O(H)(H)   or water

3 Hydrogen + nitrogen ⟶ N(H)(H)(H)   or ammonia

Carbon + 2 oxygen ⟶ O=C=O

1-9

Sample reactions between the atoms of the early earth. These compounds were probably among the first formed. (Each single line between atoms represents a single covalent bond.)

The Origin of Life and the Evolution of the Cell [13]

to such a point that its particles essentially lose all translational movement. The molecules have so little kinetic energy that the intermolecular attractive forces are strong enough to bind them into a rigid structure. The molecules are merely able to vibrate about fixed positions within this structure. Matter in the solid state, therefore, has its own form.

Most compounds can exist in all three physical states depending on the temperature. Within the range of temperatures easily attainable on the earth today, water can exist as solid ice, liquid water, and gaseous water vapor. The temperature at which a substance undergoes these changes of state is, of course, unique to that substance since it depends on two competing factors: the kinetic energy content of the molecules of the substance and their intermolecular attractions. The kinetic energy is a function of the weight of the molecules of that substance, and the intermolecular attraction depends on the electronic configuration of these molecules.

As the early earth continued to cool, some substances condensed to liquids, some even to solids, whereas others remained gases. Chronologically what probably happened was that some liquids condensed from the gaseous mass. This system continued to lose heat, especially from the surface, so a solid crust probably formed around a central molten core. A layer of gas still surrounded this system.

These conditions must have led to great turbulence. As the water vapor in the surrounding atmosphere cooled, it condensed and fell to the earth as rain. The earth's crust, however, was still very hot. When the rain fell, the heat of the solid mass caused it to revaporize immediately. Great torrents of rain fell and stormy conditions prevailed. This rain eventually formed great oceans in the low spots on the earth's crust. The rain brought compounds from the atmosphere down into the oceans where they dissolved. As the rain water flowed over the hot crust on its way to these lower oceans, it dissolved compounds from the crust as well and brought them into the oceans.

Level of Organization III: The Complex Molecule

The oceans, then, became rich in dissolved compounds and elements. Complex molecules arose from the combinations of the atoms and simple molecules that had accumulated in the waters of these prehistoric oceans and in the gases above them. A "complex" molecule is used here to designate a molecule containing any number of atoms from 10 to 200.

The common simple compounds that are believed to have been present in the gases above the oceans are methane (CH_4), carbon dioxide (CO_2), ammonia (NH_3), and hydrogen (H_2). Under the conditions on earth today these compounds are relatively inert. That is, they do not interact with one another to form complex molecules. The conditions on this early earth, however, were far different than those of today. The extreme turbulence on the earth caused by the stormy atmosphere was accompanied, of course, by electrical discharge such as lightning. In addition, the earth was also bombarded with a great deal of intense radiation from the sun. Substances such as ozone, which are currently in the atmosphere above the earth, filter out this radiation from the sun. In these prehistoric times, however, these substances were not present to shield the earth. The radiation which hit the earth was more like x-ray or atomic radiation than the light radiation that presently bathes the earth. Because of this electrical discharge and powerful solar radiation, vast quantities of energy were striking the earth. These energy sources could have provided energy to drive the synthesis of complex molecules from these simple molecules.

Laboratory experiments have shown that, if electric sparks are passed through a glass vessel containing NH_3, CH_4, H_2O, and H_2, complex molecules composed of carbon, hydrogen, oxygen, and nitrogen are formed. In fact, some of the compounds that formed, called amino acids and fatty acids, are components of present-day living things. These experiments lend credence to the theory that biologically important complex molecules arose from the interactions of these simple compounds under the conditions of the early earth. Compounds composed primarily of carbon, hydrogen, oxygen, and nitrogen are called organic compounds because most of the compounds formed in living organisms are composed of these atoms.

The Important Role of Carbon

Of the atoms carbon, hydrogen, nitrogen, and oxygen which were present in the simple molecules CH_4, NH_3, CO_2, H_2O, and H_2 on the early earth surface, carbon can bond to the most atoms. Carbon, $^{12}_{6}C$, has four electrons in its outer electron shell. It therefore needs four more electrons to fill this outer shell. Each carbon atom can bond covalently to four other atoms. In methane, CH_4, carbon is bonded covalently to four hydrogen atoms (Fig. 1-10). The atoms hydrogen, oxygen, and nitrogen need only one, two, and three

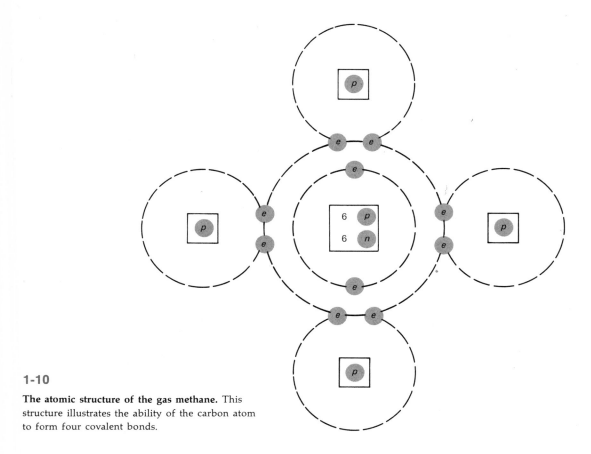

1-10

The atomic structure of the gas methane. This structure illustrates the ability of the carbon atom to form four covalent bonds.

electrons, respectively, to fill their outer shells. Carbon can form relatively stable covalent bonds with the other atoms (H, O and N) which were present as well as with other carbon atoms. Since each carbon atom could bond to four other atoms, either hydrogen, carbon, nitrogen, or oxygen, the number of different possible molecular combinations that these atoms could have formed is very large. A wide variety of compounds did form on this early earth. Those compounds that formed were basically chains of carbon atoms with other atoms of carbon, hydrogen, nitrogen, and oxygen bonded to these chains (Fig. 1-11). In some instances one end of

1-11 **The carbon chain.** The carbon atom can bond to itself to form chains with open bonds for attachment of other atoms or groups of atoms.

C   represents one carbon atom

—— represents one covalent or potential (open) covalent bond

[16]   The Origin of Life and the Cell

a carbon chain even bonded to the other end to form a molecule composed of a ring of atoms.

The chemical characteristics of these compounds will, of course, result from how these atoms are arranged within a particular molecule. For instance, a molecule composed of C_2H_6O will have different characteristics depending on how the atoms are bonded. If the oxygen atom is between the two carbon atoms, the compound is an ether, which is a gas at room temperature. If, instead, the oxygen atom is bonded to the two-carbon chain, the compound is an alcohol, which is a liquid at room temperature (Fig. 1-12).

Important Complex Molecules

The molecules that developed from the interactions between CH_4, NH_3, CO_2, and H_2O in the presence of an energy source dissolved in the prehistoric oceans. The compounds which probably formed were simple sugars, glycerine, fatty acids, amino acids, nitrogen bases, and nucleotides (Fig. 1-13). These compounds are all building blocks of the substances found in living things today. The chemical structures of these compounds will be briefly described to indicate the kinds of atomic combinations that did form and to provide a basis for understanding the further reactions which these compounds underwent at later stages in the earth's development. The chemical structure and characteristics of these compounds are described more fully in Chapter 3.

SIMPLE SUGARS: Sugars are compounds in which the atoms carbon, hydrogen, and oxygen have combined in a ratio of 1:2:1. They are generally chains of either five or six carbon atoms in which most of the carbon atoms are covalently bonded to both a hydrogen

1-12

The chemical properties of a substance depend on the arrangement of the atoms within a molecule of that substance. These two molecules, composed of the same atoms (C_2H_6O) in different arrays, have quite different characteristics.

An ether (gas)

An alcohol (liquid)

CH_4

$CH_4 + H_2O$ —energy input→ Sugars, Glycerine, Fatty acids, Amino acids, Nitrogen bases

NH_3

1-13

A possible pathway for the early formation of complex molecules important to the later formation of living things.

The Origin of Life and the Evolution of the Cell [17]

and an oxygen in addition to the two carbons in the carbon chain. The oxygen atoms are generally bonded to one carbon atom and one hydrogen atom. Different sugars were formed because each carbon or oxygen atom did not always bond in this fashion. Important five-carbon sugars (C_5) that probably developed were ribose (Fig. 1-14) and deoxyribose (see Fig. 3-10). Sugars with a C_6 backbone are glucose and fructose (Fig. 1-14).

GLYCERINE: This is a simple molecule similar to sugars except that it is only a three-carbon chain. Each carbon of the chain is bonded to a hydrogen and a hydroxyl (—OH) group as well as to the carbons of the chain. The fourth bond of the two end carbons is to hydrogen (Fig. 1-14). Glycerine is found in the fats of animal cells.

FATTY ACIDS: These compounds are composed of long chains of carbon atoms of different lengths. Aside from the bonds of the carbon chain, all the carbon atoms are bonded entirely to hydrogen atoms except the carbon at one end of the chain. This carbon is bonded to two oxygen atoms (Fig. 1-14). This carboxyl (—COOH) function makes these compounds slightly acidic. Fatty acids are found combined with glycerine in the fat of animal cells.

AMINO ACIDS: There are many different kinds of amino acids. They vary in both the length of their carbon-chain backbones and the atoms connected to this backbone. In all amino acids, however, one end carbon is bonded to two oxygens to form the acidic —COOH group found also in fatty acids. In addition, the carbon next to this group is bonded to a nitrogen atom in all the amino acids. The nitrogen atom, along with its two covalently bonded hydrogens, is called an amino group (—NH_2). These varied compounds with both an amino and an acidic group are called amino acids (Fig. 1-14).

NITROGEN BASES: These compounds are composed of complex ring structures containing carbon and nitrogen atoms. There are two types of these bases. One type, the pyrimidines, is composed of a single ring of six atoms. The other type, the purines, possesses a fused double ring composed of nine atoms (Fig. 1-14). The different bases of each type vary in the atoms or groups of atoms bonded to these basic ring structures.

NUCLEOTIDES: Nucleotides are substances that are somewhat more complex than the compounds already discussed. They are formed by a combination of two of these compounds, a sugar and a nitrogen base. In addition, a nucleotide also contains an inorganic phosphate group bonded to the sugar (Fig. 1-14). Since there are several

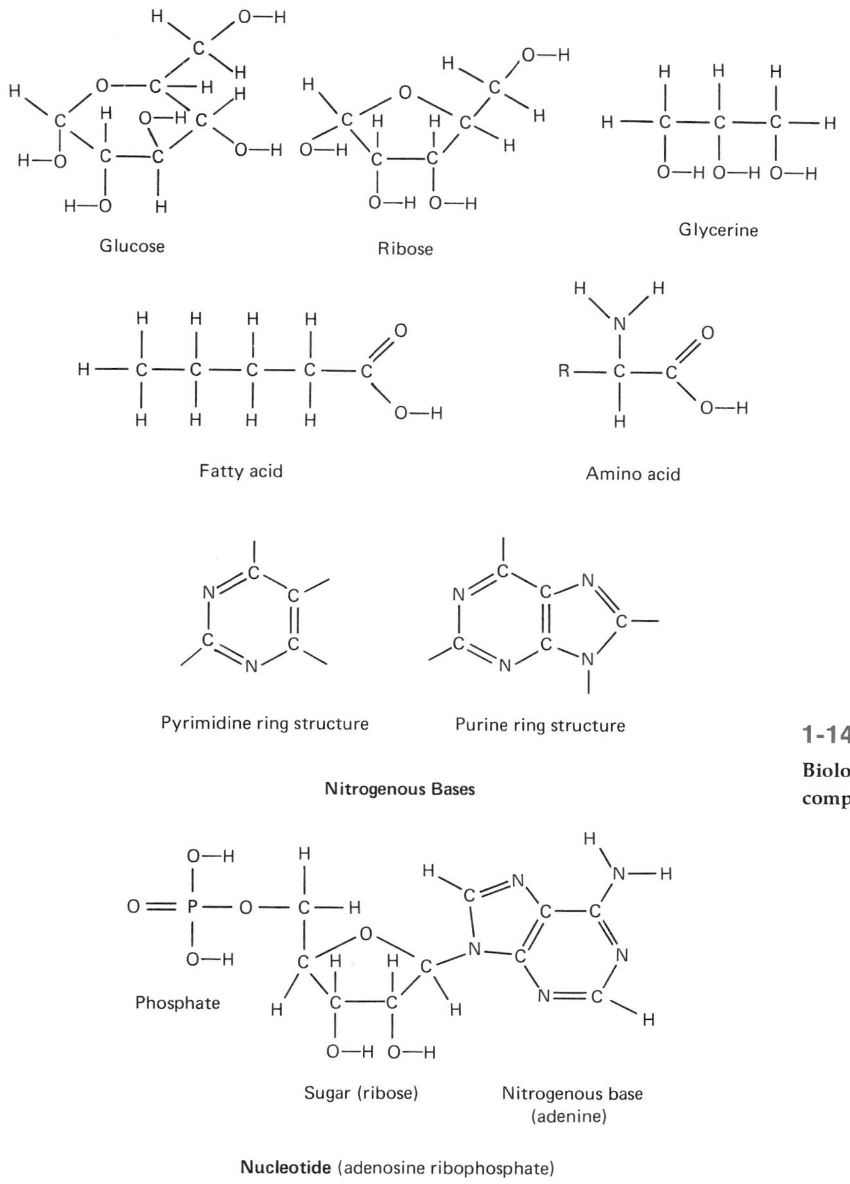

1-14
Biologically important complex molecules.

different sugars and nitrogen bases, nucleotides can differ in either the sugar or the nitrogen base. They all have the same basic structure of base-sugar-phosphate, however. The nucleotides which are apparently important in the development of living organisms contained the C_5 sugars, ribose and deoxyribose.

The Origin of Life and the Evolution of the Cell [19]

Level of Organization IV: The Macromolecule

Macromolecules, as their name implies, are very large molecules. In this text they are those molecules which have more than 200 atoms. Many macromolecules found in living things have a great deal more than 200 atoms. Most of the macromolecules which developed from that rich organic soup of the early oceans were polymers formed from the complex molecules already discussed. A polymer is a very long chain of atoms built up by linking together, one after another, identical or similar compounds, called monomers. Each link between these smaller compounds is always the same type of chemical bond. That is, linkage occurs between the same two kinds of atoms. A monomer, then, has two reactive centers. One of these centers can react with the other center to form a bond, but it only reacts with that center of a different molecule. For instance, if atom A can bond to atom B as in the reaction RB + R'A → RB—AR', then a molecule containing both atoms A and B, ARB, can react with other such molecules to form a long chain ARB—ARB—ARB—ARB—ARB—. . . .

Sugars have many reactive centers since each carbon is bonded to a much more reactive oxygen atom. Therefore, sugars can link together to form long chains in different ways. Both starch and cellulose are different polymers of the C_6 sugar glucose (Fig. 1-15). Sugars and polymers composed of sugars are called carbohydrates.

Amino acids have two reactive sites, the acid group and the amino group. These two groups can bond together to form long chains

1-15 Common examples of polymer macromolecules. All are formed by the connection of many similar molecules into a long chain.

Basic Structure of Starch — A Polymer

· · · Glucose — Glucose — Glucose — Glucose — Glucose — Glucose · · ·
Molecule   Molecule   Molecule   Molecule   Molecule   Molecule

Basic Structure of a Protein — A Polymer

· · · Amino — Amino — Amino — Amino — Amino — Amino — Amino — Amino — Amino · · ·
Acid   Acid   Acid   Acid   Acid   Acid   Acid   Acid   Acid

Basic Structure of a Nucleic Acid — A polymer

· · · Nucleotide — Nucleotide — Nucleotide — Nucleotide — Nucleotide · · ·

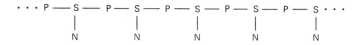

1-16 The basic structure of the nucleic acid. The phosphate and sugar units are joined by covalent bonding to form a long chain of atoms, and the nitrogenous bases form side groups.

P represents a phosphate group

S represents a sugar molecule

N represents a nitrogenous base

of amino acid residues. A polymer chain composed of amino acids linked together is a protein (Fig. 1-15).

Nucleotides bond together between the phosphate group of one and the sugar of another to form long-chain molecules called nucleic acids (Fig. 1-15). In the nucleic acids, the nitrogen bases of the nucleotide are not involved in the polymer linkage. These nitrogen bases hang off this sugar-phosphate-sugar- . . . polymer backbone (Fig. 1–16). Molecules of this type are sometimes thousands of nucleotides long.

Level of Organization V: The Cell

In time the primordial oceans were vast mixtures of compounds of various sizes and types. They contained macromolecular polymers, such as nucleic acids, proteins, and carbohydrates; various other complex, but smaller, molecules, such as sugars, amino acids, nucleotides, fatty acids, and nitrogen bases; as well as the simple molecules such as NH_3, CO_2, CH_4, and H_2. Organic compounds, especially macromolecules, have more affinity for one another than they do for water. Because they were more attracted to one another than to the water in which they were dissolved, these compounds are believed to have aggregated together.

There are several theories as to where such aggregation occurred. It could have occurred in pockets along the ocean shore, on rocks or other projections protruding from the ocean floor, or on clay or other solid particles suspended within the waters themselves. It has been shown experimentally that such organic compounds aggregate together to form droplets called coacervates within a water medium. Although coacervates have a film-like covering around them which serves to separate them somewhat from the water medium, some compounds can still flow between the coacervates and the surrounding medium. Coacervates have been shown to form in solutions of organic macromolecules which are approximately 100 times more dilute than the primeval oceans are believed to have been. It is very probable, therefore, that these molecules did form aggregates.

All the possible ways that such chemical aggregates might have formed in the prehistoric oceans involve a random accumulation of compounds. The aggregates that formed, then, all differed from one another in composition. The molecular components were in such close proximity within these aggregates that they could interact with one another. These interactions might be "weak" and merely serve to organize compounds relative to one another, or they might be "strong" and result in the formation of different compounds. Some of the different compounds that formed due to these strong interactions might not be held tightly to the aggregate and would go off into the medium. Depending on their components some aggregates would attract specific compounds from the medium more than other aggregates would. The aggregates, therefore, were in a very dynamic state. Some aggregates would be more stable than others, depending on the molecular components and their interactions and the exchange of compounds that occurred between the aggregates and the medium. Only the aggregates which had enough attractive forces to offset the disruptive forces would last.

Prior to the formation of aggregates, reactions between organic compounds probably occurred at random, depending upon the chance encounters of molecules within the primeval broth. One molecule might possibly undergo several different reactions depending upon the other molecules it encountered (Fig. 1-17). Once aggregates of organic compounds of various sizes and types formed, these component molecules probably underwent reactions in a less haphazard manner. There are many sites along the length of the organic macromolecules which could attract other molecules of the aggregates. If some molecules were weakly bonded to neighboring sites along the same macromolecule, these molecules might be close enough to one another to interact (Fig. 1-17). A specific interaction brought about in this manner would have a greater chance of occurring in such aggregates than it would if the reactant molecules were floating around freely in the dilute primitive oceans.

Since many of the types of molecules present in these primitive aggregates are the same as those found in the cells of present-day organisms, it seems reasonable to assume that the processes which occur between these molecules in cells might also have occurred

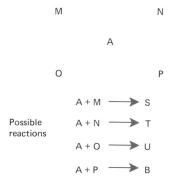

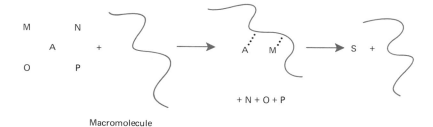

1-17

A possible mechanism which could have occurred within aggregates to bring about specific reactions.

in these primitive aggregates at some point in their development. In cells of currently living organisms both nucleic acids and proteins play a major role in "directing" the cell processes. Both these macromolecules exert their influence basically by the mechanism outlined above. Nucleic acids apparently direct the synthesis of specific sequences of monomers into the macromolecular polymers. Proteins influence other specific reactions of the cell.

The protein and nucleic acid molecules which were originally incorporated in the primitive aggregates probably formed during random collisions of monomers with growing polymer chains which occurred within the ocean waters. Not all the monomers of either proteins or nucleic acids are identical. At each linkage in the formation of these macromolecules, any one of several possible monomers could be incorporated. Since these molecules grew to such lengths with possible variation at every step of the way, it is very unlikely that any two molecules of either protein or nucleic acid formed by such random combinations would be identical.

The Origin of Life and the Evolution of the Cell [23]

Once chemical aggregates had formed which included these macromolecules, however, additional macromolecules could have been synthesized by quite a different mechanism. For instance, if each type of monomer that was already incorporated into an existing macromolecule bonded weakly to a specific unincorporated monomer, these weak interactions would serve to "line up" a specific array of monomers. Polymer formation between monomers which are in such close proximity to one another would be far more probable than between free-floating monomers that only collide by chance. Some aggregates probably developed that had a mechanism of this kind that allowed the synthesis of a specific macromolecule. If so, that synthesis would probably occur faster than a random synthesis since the monomers, due to their chemical interactions, would be placed in proximity to one another and would not have to "wait" for chance encounter. Hence, once aggregates formed that were stable, the greater percentage of the macromolecules that were synthesized within them probably had a specific structure that was somehow related to the structure of other macromolecules within the aggregates.

The macromolecules with specific structure, could only interact with certain other molecules. By virtue of these specific interactions between macromolecules and other molecules, only certain reactions would occur in the aggregates. Aggregates then, became, by the nature of their components and interactions, able to perform certain chemical functions. For instance, if molecule A encountered an aggregate with certain components, it was always transformed into molecule B. As aggregates became more complex, they probably could perform several functions: A $\longrightarrow$ B $\longrightarrow$ C $\longrightarrow$ D $\longrightarrow$ E and so forth.

The synthesis of macromolecules within aggregates could continue as long as the supply of monomers lasted. Aggregates competed for monomers from the medium. The chemical functions that some aggregates could perform might have been to build some of these monomers from even simpler compounds. Aggregates that could do this would continue to synthesize macromolecules which in turn "directed" chemical function. The aggregates that were most "successful" were those that could maintain their own structure and function. These aggregates were stable, the others probably gradually redissolved as they lost compounds to the medium.

When aggregates evolved that could completely maintain their own structure and function, merely from interacting with the simple compounds in their environment, a primitive living system had developed. Judging from cells of current organisms, these aggre-

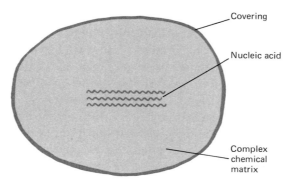

1-18 The form of the first cell?

gates probably had a nucleic acid core that was surrounded by some sort of matrix of proteins and other organic molecules where chemical reactions occurred. This system was encased by some sort of protective covering (Fig. 1-18). The nucleic acids were the macromolecules which ultimately directed the synthesis of all other macromolecules. Aggregates developed in which the nucleic acids were able to direct the synthesis of exact duplicates of the nucleic acids. When these aggregates got to be too big and split, the two halves, both containing identical nucleic acids, developed the same structures and functions as the original aggregate. In this way aggregates with this capability would last and "reproduce."

Many of the chemical functions that these early life forms exhibited were energy-requiring steps. In order to carry out these steps, other functions of the same living system must have been energy-releasing steps. The energy given off in the releasing steps drove the energy-requiring reactions. The energy-releasing steps were probably those of the breakdown of energy-rich organic compounds which were present in the primeval seas. Organisms which require an intake of such compounds are called heterotrophs. As the supply of these compounds dwindled, the primitive life forms that required them "failed" as they could no longer maintain their own structure and function. Some life forms must have been able to synthesize these compounds within themselves from readily available compounds. Living systems with this capability would be the most successful. To do this, these forms had to have a mechanism for trapping energy to use in forming the energy-rich compounds.

There are two mechanisms by which present-day organisms are able to trap energy in the formation of organic compounds. Orga-

Establishment of Life

nisms which can do this are called autotrophs. Some organisms are able to synthesize organic compounds from energy-rich simple compounds. This process is called chemosynthesis. An example of a present-day chemosynthetic organism is the bacterium, *Clostridium aceticum*. This organism can synthesize a two-carbon compound, acetic acid, from CO_2 and H_2 in an energy-releasing reaction (Fig. 1-19). There are relatively few such organisms with this capability, however. This indicates that life forms with this energy-trapping mechanism were not the most successful.

At some point a primitive life form developed that could trap light energy from the sun and use it to drive the synthesis of energy-rich organic compounds within itself from the readily available compounds CO_2 and H_2O. These organisms are the ancestors of the green plants, and this process which developed is called photosynthesis. A basic picture of the chemical reaction involved is displayed in Fig. 1-19. This process differs from chemosynthesis since the energy used in the synthesis of organic compounds comes from sunlight rather than from the chemical bonds of certain nonorganic molecules. Once organisms developed with the capability of using the sun's energy, they were assured of a chemical energy source to maintain their structure and function.

1-19 The difference of energy flow by chemosynthesis and photosynthesis. Chemosynthesis has an energy input of chemical bonding energy of a nonfood material. Photosynthesis has an energy input in the form of light energy. Both form food materials from energy.

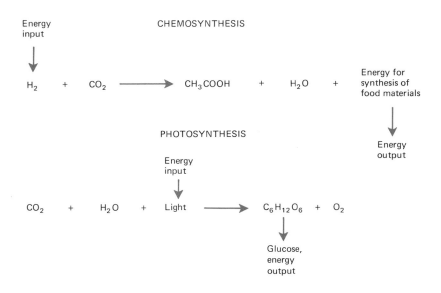

[26]   The Origin of Life and the Cell

Photosynthetic organisms produced O_2, gaseous oxygen. This oxygen reacted with the other gases present and changed the composition of the atmosphere. These reactions changed the chemical conditions on the earth and, therefore, comprise the "oxygen revolution." Oxygen reacts with CH_4 to form CO_2, with NH_3 to form N_2 gas, and with itself to form O_3, ozone gas (Fig. 1-20). The disappearance of CH_4 and NH_3 and the appearance of N_2, O_2, and O_3 changed the conditions in which organisms were living quite drastically. The simple compounds, CH_4 and NH_3, were necessary to the original formation of organic compounds which ultimately led to primitive life forms and served as their energy source. The unshielded radiant energy from the sun (containing much dangerous ultraviolet radiation) was also necessary for this formation of organic compounds. The ozone formed from the O_2 collected in the upper atmosphere and absorbed this cosmic radiation so that it no longer hit the earth. The organic compounds which were components of primitive life forms and those that were their energy sources could no longer be produced by the same mechanisms. The raw materials and energy source had essentially disappeared. All the primitive life forms that survived the oxygen revolution became dependent on autotrophs that could synthesize organic compounds within themselves from these new environmental components.

On the other hand, these developing organisms were now shielded from disruptive radiation. The O_2 building up in the atmosphere also interacted with the primitive life forms. As the chemical environment changed, the chemical functions that life forms could perform also changed. In some cases these new chemical functions provided the primitive life forms with more energy than they could attain in the earlier environment.

The appearance of photosynthetic organisms changed the environment in which the primitive life forms lived considerably. The

Oxygen Revolution

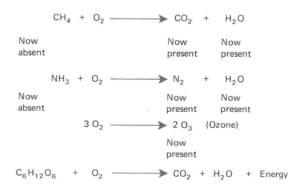

1-20

Sample reactions of the oxygen revolution. These reactions changed the atmosphere and energy cycles of the earth and made them more favorable to life as we know it.

The Origin of Life and the Evolution of the Cell

processes that led to these forms were no longer possible in this photosynthetic environment. Some of the life forms in existence, however, could continue to exist in this altered environment. It was these forms that developed into the forms of life as they are known today.

Suggested Readings

FOX, S. W., K. HARADA, G. KRAMPITZ, and G. MUELLER. "Chemical Origins of Cells." *Chem. Eng. News,* **48**(26), 80–94 (1970).

GAMOV, G. "The Origin and Evolution of the Earth," *Amer. Sci.,* **39**, 393–406 (1951).

OPARIN, A. I. *Life, Its Nature, Origin, and Development,* Ann Synge (trans.), Academic Press, New York, 1966.

UREY, H. "The Origin of the Earth," *Sci. Amer.,* **187**, 53–60 (1952).

WALD, G. "The Origin of Life," *Sci. Amer.,* **191**, 44–53 (1954).

Cell Structure 2

The description of the origin of life presented in Chapter 1 is based on the assumption that the cell is the basic unit of life. This means that a collection of compounds only possesses life if these compounds are *organized* into certain structures which are now called cells. This assumption that all living things are composed of cells, called the cell theory, was first proposed in the 1830s by Schleiden and Schwann. These men did not discover cells by any means. Cells had been observed in many organisms since 1665 when Robert Hooke first viewed them in cork using his newly invented microscope. The theory that Schleiden and Schwann proposed, however, connected these many different observations into a unified picture as to the composition of living things.

As microscopic techniques improved and cells could be more closely scrutinized, the cell looked even more like the basic unit of life. Under high magnification cells exhibit a specific organization of their components. Cells from a wide variety of organisms were all found to possess the same general structural organization and many similar internal structures. A cell appeared to be a watery matrix of organic and inorganic components surrounded by a filmlike covering or membrane. Within that water matrix, called the cytoplasm, some of the cellular components were organized into specific structures. These internal structures are called organelles, and some organelles are common to most cells. The nucleus is the central body in the cytoplasm in most cells. Other organelles found in most cells are mitochondria, the endoplasmic reticulum, ribosomes, the Golgi apparatus, and lysosomes. Some cells with more specialized functions possess other organelles such as chloroplasts, a centriole, and flagella and cilia. Presumably the specific organization of components into these organelles is responsible for the cell's ability to perform various functions.

Cells do vary in size and structure from organism to organism and even within a many-celled organism. Most cells are too small to be seen by the unaided eye. The human eye cannot resolve things that are less than 100 microns apart. That is, any two units which are closer together than 100 microns appear as one unit to the eye. Due to their size cells are measured most conveniently in terms of microns (μ), millimicrons (mμ), and Angstroms (Å).

1 inch (in.) = 25,400 microns (μ) = 25.4 millimeters (mm)
1 μ = 1000 millimicrons (mμ) = 10,000 Angstroms (Å) = 0.001 mm

Light microscopes can resolve things up to 200 mμ apart. This resolution is about 500 times better than that of the unaided eye. Many of the cells that make up the tissues of multicellular plants and animals are about 20 μ wide on the average. These cells and even some of their internal structures can be seen with a light microscope.

Many bacterial cells, all viruses, and many of the structures within cells are too small to be seen even with the light microscope. They can be "seen" with the electron microscope, however. This microscope has a resolution of about 5 Å. Since the recent advent of electron microscopy, the details of cell structure have been quite extensively studied. Electron microscopy has revealed some structures that were previously unknown. Unfortunately, because of the nature of the electron beam which is responsible for the high resolving power of this microscope, only samples that are exceedingly thin can be "looked at" with an electron microscope. Only cells that are specially prepared and no longer alive can be studied with this scope. There is always the chance that cell structures found may be a direct result of the method of preparing the sample for "viewing" and may not be typical of the living cell. Electron microscopists must make sure that their findings are not such artifacts.

As the "optical" methods for viewing cell structures improved, the chemical and physical methods for identifying cell components and studying cell function also improved. In recent years a great deal of information has been amassed about the composition, structure, and function of a wide variety of cells. Cells from all organisms have a great deal in common. They are composed of similar macromolecules which are built up in the same way of the same basic building blocks. These similar components are organized into many of the same kinds of organelles. Finally, cells perform basically many of the same functions. Some functions have been found to occur within certain organelles. A brief description of cellular structures and their probable roles in the various cell processes is presented in this chapter. Unit II is devoted to a more detailed description of cell function. Certainly the unity of cell design, cell function, and cell composition gives plausibility to chemical evolutionary theory as well as cell theory.

Intracellular Structure

Cell Membrane

All cells are surrounded by a cell membrane. Cell membranes, under the high magnification of an electron microscope, appear to be composed of three layers. The chemical composition of the two outer layers is thought to be protein and the inside "filling" layer, modified fats. The structure of the cell membrane, however, is not known with certainty as yet. The whole membrane is only 80–100 Å thick.

Ground Cytoplasm

The main component of cytoplasm is water. This water, however, has many different kinds of chemical compounds in it. These compounds are the nucleic acids, enzymes, proteins, and various other chemicals that are involved in the life functions of the cell. Some of these compounds are in solution in the water; others are in a colloidal suspension. In both solutions and colloids, substances are uniformly distributed throughout the water medium. In a solution, however, individual molecules or ions of a substance are distributed throughout the medium. In a colloid, clumps of molecules rather than individual molecules are distributed among the water molecules. These clumps are generally about 1–100 mμ in size.

The factor that determines whether a substance forms a solution or suspension in water is the electronic interaction between the molecules of the substance and the water molecules. Because the oxygen atom in H_2O has a stronger attraction for the electrons in the H-O bond than the hydrogen atom does, the oxygen atom is slightly negative and the hydrogen atoms are slightly positive. A molecule with such an unequal electron distribution of bonding electrons is said to be "polar". Because of this polarity of water, other polar compounds, especially ionic compounds, are readily soluble in water. The polarity of the water molecules lessens the electrostatic forces holding an ionic crystal together so that it breaks apart more readily and dissolves. Once the ions are in solution, the polarity of the water neutralizes the effect they have on one another, and the solution is stable (Fig. 2-1).

Some compounds are made up of atoms which have about equal attraction for the bonding electrons that hold them together. Therefore, the bonding electrons are not attracted to one atom of the bond more than the other. Compounds with such unpolarized bonds are considered "nonpolar" and do not dissolve readily in water. The polarity of water does not lessen the intermolecular forces holding a nonpolar compound together. Many organic compounds with many C-H bonds are basically nonpolar, since the C-H bond is not polarized. Such molecules have greater affinity for molecules like themselves than they have for water. In the extreme case, molecules that are composed entirely of carbon and hydrogen form a separate liquid layer which does not mix with

2-1

The dissolving of table salt in water and the formation of a solution of salt and water. The pull of the water molecules helps the sodium and chloride ions break away from the salt (a) and disperse (b). After these ions disperse they are held permanently in solution by their interaction with the water molecules (b).

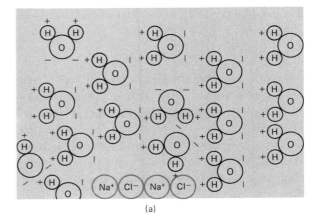

(a)

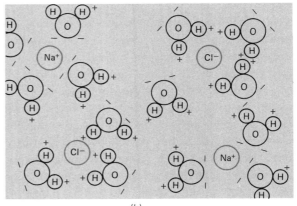

(b)

[32] The Origin of Life and the Cell

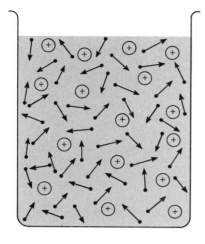

2-2

Forces that can keep colloidal particles in permanent suspension in a colloid. The like charges of the particles repel each other, and the water molecules buffet the colloid particles in their high-velocity and random movement.

⊕ represents a colloidal particle with positive charge.

•→ represents a water molecule in high velocity, random, movement

water. However, small groups of nonpolar molecules can be "suspended" in water if these groups form around a molecule which is largely nonpolar but which can ionize. The groups of molecules which form around ionizable molecules of the same substance are all identically charged. Therefore, they mutually repel each other and distribute themselves throughout the fluid, forming a colloid (Fig. 2-2).

Some cellular components, such as storage granules and fat droplets are not dissolved or suspended in the cytoplasm. These compounds are products of cell function which have collected in specific spots within the cytoplasm. The cytoplasm also contains vacuoles. A vacuole is an area enclosed within the cytoplasm that is surrounded by a vacuolar membrane. This membrane is similar in structure to a cell membrane. A vacuole is generally filled with a watery mixture.

Organelles of the Cytoplasm

Nucleus The most prominent organelle in the cell is the nucleus. This is a fluid-containing body that is separated from the cytoplasm by the nuclear envelope. Many experiments with single-celled organisms in which nuclei were removed and transplanted among slightly different organisms indicate that the nucleus exerts continuous control over the cytoplasm. "Cells" without nuclei lose some functions, and cells with transplanted nuclei take on some

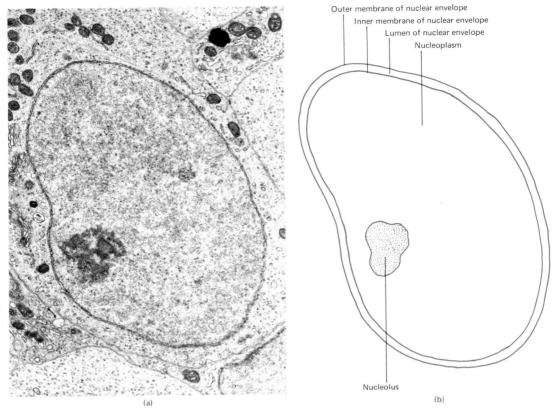

2-3 The structure of the nucleus of a cell, as shown by (a) an electron micrograph and (b) a labeled diagram. (*Courtesy of Charles J. Flickinger, M. D., University of Colorado, Department of Molecular, Cellular, and Developmental Biology.*)

of the characteristics of the cell from which the transplanted nuclei was taken.

NUCLEAR ENVELOPE: The nuclear envelope is composed of two membranes (Fig. 2-3). The outer membrane is generally continuous with the endoplasmic reticulum and often has ribosomes attached to it. Electron microscopy indicates that there are pores about 500-Å long which interrupt the nuclear envelope (Fig. 2-4). Whether or not substances pass freely between the nucleus and the cytoplasm through these pores is not known at this time. Some microscopists believe that there is a fine partition across the pores of the nuclear membrane which might hinder such free transport.

NUCLEOPLASM: The fluid medium of the nucleus is called the nucleoplasm. It is a colloidal suspension which contains proteins, the nucleic acids DNA (deoxyribonucleic acid) and RNA (ribonucleic acid), and other chemicals of the nucleus. The many chemical reactions of the nucleus which are extremely important to cell function occur in this medium.

CHROMATIN: If a cell is specially stained, fine threads of material are evident within the nucleus. This material, composed of proteins and DNA is called chromatin. When a cell divides, this material condenses into thicker rodlike structures referred to as chromosomes.

The DNA portion of chromatin has a major role in directing cell function. The mechanisms by which DNA does this will be discussed more fully in Chapters 4 and 5.

NUCLEOLUS: The nucleolus is a spherical particle within the nucleus which does not have a covering membrane (Fig. 2-3). There may be one or more nucleoli in a single nucleus. The nucleolus is composed primarily of DNA, RNA, and proteins. This body is involved in the process of protein synthesis, since the nucleic acid component of ribosomes is synthesized there.

Mitochondria Chemical reactions which take place in the mitochondria give the cell a source of "useable" energy. Because of this function, mitochondria have come to be known as the powerhouses of the cell. Most cells contain mitochondria.

A mitochondrion is a small oblong-shaped body generally 1–10 μ long and 1–2 μ in diameter. If viewed through an ordinary microscope, they appear only as small granules within the cytoplasm. At the high

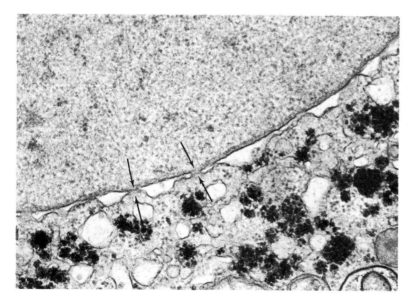

2-4

The pores of the nuclear envelope as shown by an electron micrograph of the nuclear envelope (see arrows). (*Courtesy of A. E. Vatter, University of Colorado, Webb Waring Institute.*)

2-5

The structure of the mitochondria as given by (a) an electron micrograph and (b) a labeled diagram. (*Courtesy of Charles J. Flickinger, M. D., University of Colorado, Department of Molecular, Cellular, and Developmental Biology.*)

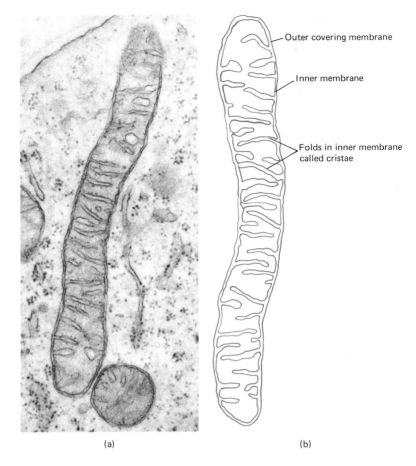

(a)   (b)

magnification provided by electron microscopy, however, a mitochondrion appears to have a double membrane around it. The outer membrane is merely a single layer enclosing the mitochondrion. The inner membrane, however, has many convolutions extending into the organelle. These convolutions of folds are called cristae (Fig. 2-5). Some cristae extend across the width of the mitochondrion and some only extend partially into the interior.

Most of the energy-producing reactions which occur in the mitochondria take place on the surface of these cristae. Mitochondria in cells with high energy requirements, such a muscle cells, typically have many cristae. Cells with low energy requirements, however, have mitochondria with fewer cristae.

Lysosomes Lysosomes are small bodies in the cytoplasm which contain enzymes that enhance the breakdown of cellular

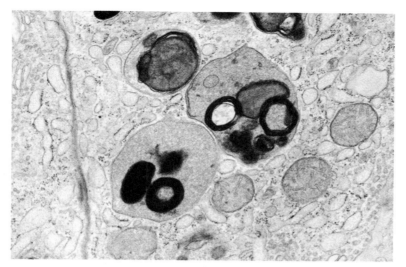

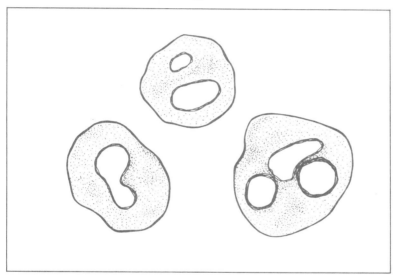

2-6

The structure of lysosomes as given by (a) an electron micrograph and (b) a labeled diagram. The bodies within the lysosomes are apparently being broken down. (*Courtesy of Charles J. Flickinger, M. D., University of Colorado, Department of Molecular, Cellular, and Developmental Biology.*)

(a)

(b)

components. The structure and size of lysosomes vary, but they are generally spherical and about 500 mμ in diameter. They are covered by a single membrane (Fig. 2-6). The functions of lysosomes are not known precisely, but apparently they release the digestive enzymes which they contain both into vacuoles containing food particles and into damaged cytoplasmic organelles to cause the breakdown of their components. Lysosomes may also be used to break down the components of entire cells which are no longer useful or have died.

Cell Structure [37]

2-7

The form of the granular endoplasmic reticulum as given by (a) an electron micrograph and (b) a labeled diagram.
(Courtesy of Charles J. Flickinger, M. D., University of Colorado, Department of Molecular, Cellular, and Developmental Biology.)

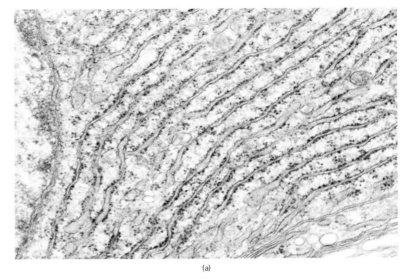

(a)

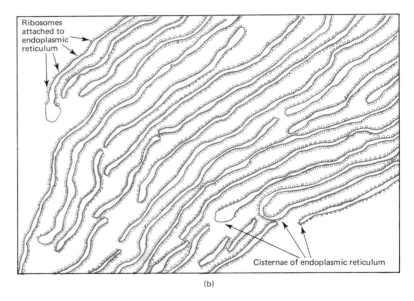

(b)

Endoplasmic Reticulum The endoplasmic reticulum is a complex system of membranes which forms a collection of membrane-bound cavities; these often interconnect into a membrane-bound system of channels within the cytoplasm. The size and shape of these cavities vary with the type of cell. When the cavities are saclike, they are called cisternae. The endoplasmic reticulum is often attached to the nuclear membrane but not to the cell membrane. This cellular structure was not noted until the advent of

[38] The Origin of Life and the Cell

electron microscopy, and then two types of endoplasmic reticulum were observed: the rough, or granular, and the smooth, or agranular, types. The granules attached to the granular endoplasmic reticulum (which are responsible for its name) are ribosomes.

The pattern formed by the membrane-bound cavities of the granular endoplasmic reticulum varies from a few isolated cisternae to extensively interconnected systems. It sometimes appears as a system of tubules and sometimes as a system of flattened sacs. These flattened cisternae are often found to be parallel to one another, giving a layered appearance to the granular endoplasmic reticulum (Fig. 2-7).

The granular endoplasmic reticulum apparently serves several cell functions. Because of the attached ribosomes, it can be the site of protein synthesis. Proteins which ultimately will be secreted by the cell are synthesized there. The vesicles and cavities of the endoplasmic reticulum probably serve in the segregation and transport of these proteins in preparation for further processing and discharge from the cell. This endoplasmic reticulum may be involved in the collection or "packaging" of digestive enzymes to form lysosomes.

Sometimes the agranular form of endoplasmic reticulum is attached to the granular form, but it differs from the granular form in both structure and function. Structurally, the agranular form is an extensive tubular system without any flat cisternae (Fig. 2-8). The agranular form is found in many different kinds of cells and is believed to be involved in several cell functions. For instance, this smooth form is quite prevalent in the cells in which sex hormones are synthesized, and it is believed to be involved in this synthesis. Its presence in the cells which line the intestine is believed to indicate that it is also used in the transportation of fats. The other kinds of cells in which it abounds suggest that the agranular form has some part in many more equally varied functions.

Golgi Apparatus The Golgi apparatus is an assembly of flat saclike cisternae which looks like a stack of saucers (Fig. 2-9). Golgi systems differ in size and compactness. A cell may have one or more such structures.

Basically, Golgi apparati seem to function as the points within the cell where compounds to be secreted by the cell are collected and concentrated. In the human pancreas, for example, enzymes synthesized on the ribosomes are collected in the membranes of the Golgi apparatus and then are secreted. When the cellular secre-

2-8

The form of agranular endoplasmic reticulum as shown by (a) an electron micrograph and (b) a labeled diagram. The granular and agranular endoplasmic reticulum are attached on this micrograph, showing the difference in appearance rather clearly. Note the "granular" appearance given to the granular endoplasmic reticulum by the ribosomes but lacking on the agranular endoplasmic reticulum. (*Courtesy of A. E. Vatter, University of Colorado, Webb Waring Institute.*)

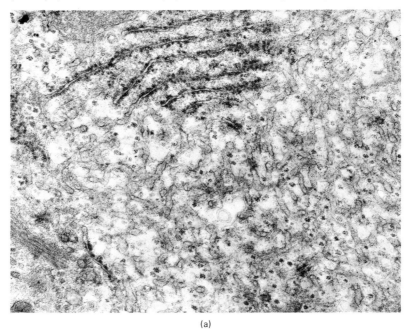

(a)

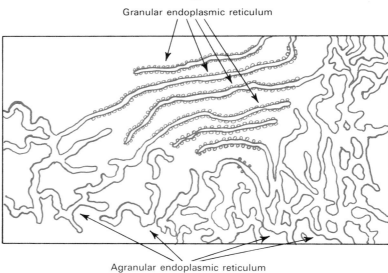

Granular endoplasmic reticulum

Agranular endoplasmic reticulum

(b)

tions are a combination of proteins and carbohydrates, the carbohydrates may be synthesized in the Golgi apparatus and the complex of carbohydrates and proteins is assembled there. Lysosomes may form at the Golgi appartus when digestive enzymes are collected there.

Ribosomes Ribosomes are small granules about 15–20 mμ in diameter. They occur distributed throughout the cytoplasm as well as attached to the endoplasmic reticulum. Unlike the other organelles already described, ribosomes are not surrounded by a membrane (Fig. 2-7). Ribosomes are composed of ribonucleic acid (RNA) and protein. These small bodies are essential to cell function since protein synthesis occurs at the ribosomes. The number of ribosomes found in a cell reflects the cell's potential for protein production.

Organelles Found Only in Certain Kinds of Cells

Chloroplasts Chloroplasts are large organelles found mainly in plant cells. They are green and are responsible for the green color which plants exhibit. All the chlorophyll of a plant is contained in its chloroplasts. These bodies are the site of photosynthesis.

Chloroplasts are large enough to be seen with a light microscope. They are enclosed by a membrane, but the internal membraneous structure of chloroplasts is quite complex. The interior contains many "stacks" of membranes called grana which are connected to one another by a different system of membranes. The grana are composed of proteins, modified fats, chlorophyll, and other pigments arranged in a layered structure. The grana are round but exhibit this layered structure in cross section (Fig. 2-10). The extensive membrane system of the grana offers a great deal of surface area for compounds to undergo the chemical reactions of photosynthesis.

Centriole Centrioles are cylindrical structures that are about 300–400 mμ long and 150 mμ in diameter. They are composed of nine sets of triplet fibers (Fig. 2-11). The inner fiber of each triplet is usually connected to the outer fiber of the adjacent triplet by a subfiber. The cells of the higher animals have two centrioles located near the nucleus. The two centrioles in an animal cell are usually at right angles to one another. Centrioles are not found in the cells of higher plants.

During cell division the centrioles replicate themselves to form four centrioles. Two centrioles move to each side of the nucleus

2-9

The form of the Golgi apparatus, as shown by (a) an electron micrograph and (b) a labeled diagram. Typically it takes the shape of a stack of membranes and is located close to the nucleus. (*Courtesy of Charles J. Flickinger, M. D., University of Colorado, Department of Molecular, Cellular, and Developmental Biology.*)

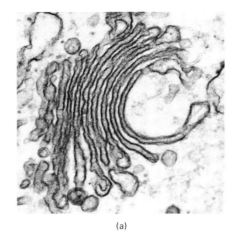

(a)

(b)

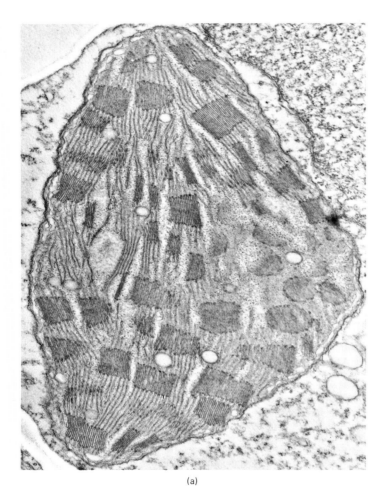

(a)

2-10

The structure of the chloroplast, as shown by (a) an electron micrograph and (b) a labeled diagram. The interior of the chloroplast is a maze of membranes, many of which are "stacked" into structures called grana. Note the layered structure of the grana in cross section, but the round form of the grana in face view. (*Courtesy of A. E. Vatter, University of Colorado, Webb Waring Institute.*)

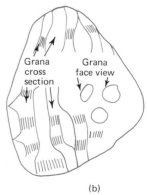

(b)

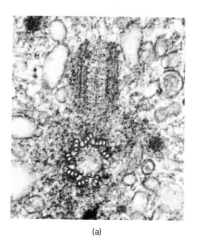 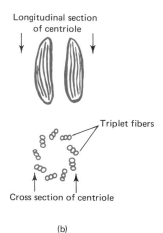

2-11

The structure of centrioles as shown by (a) an electron micrograph and (b) a labeled diagram. Note that the centrioles occur in pairs and that they often lie at right angles to each other. The basic structure of the centrioles is a cylinderlike assembly of triplet fibers. (*Courtesy of A. E. Vatter, University of Colorado, Webb Waring Institute.*)

of the dividing cell. They position themselves at cellular locations called the poles and apparently form a system of fibers that redistributes the chromosomes during cell division (Fig. 2-12). This assembly of fibers is very prominent in a dividing cell. A similar assembly of fibers is also formed in plant cells which do not contain centrioles. The precise function of the centrioles is, therefore, not well understood at this time.

Cilia and Flagella Cilia and flagella are organelles located on the cell surface. They possess threads which protrude from the cell and beat or vibrate. Single-celled organisms use these structures for locomotion through a medium. Stationary cells of multicelled organisms use them to move substances across the cell surface. Although these structures are quite similar anatomically, a flagellum is considerably longer than a cilium. A cell with flagella usually only has one, two, or four of them, but a cell with cilia has row upon row of cilia.

Externally, these structures are hairlike protrusions from the cell membrane. Internally, they are composed of nine double fibrils arranged in a cylindrical ring around two central, single fibrils (Fig. 2-13). The central fibrils terminate at a granule just above the surface of the cell. At the surface of the cell, just below this granule, there is a membrane called the basal plate. The cilium and flagellum rest on a basal body in the cytoplasm below the basal plate. This body has a cylindrical structure much like that of the centriole. It is also composed of nine sets of triplet fibrils (Fig. 2-11).

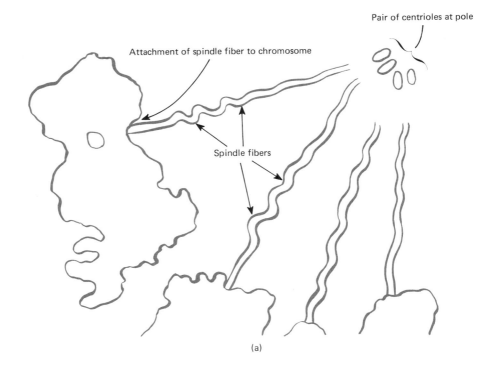

(a)

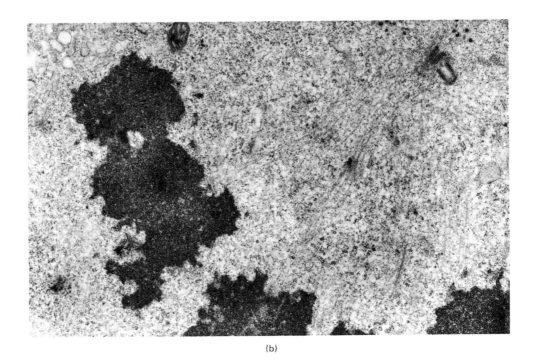

(b)

2-12 (opposite)

The involvement of the centrioles in the cell division of a cell as shown by (a) an electron micrograph and (b) a labeled diagram. The centrioles form the basis for the formation of the aster and the spindle fibers. The spindle fibers attach to the chromosome during the redistribution of chromosomes during cell division. (*Courtesy of A. E. Vatter, University of Colorado, Webb Waring Institute.*)

Extracellular Structure

Cell Wall

The cells of plants are surrounded by a rigid or semirigid covering made, in part, of cellulose. These cell coverings are called cell walls, and animal cells do not have them. The cell walls help give plants their form.

The cellulose for the cell wall is presumably synthesized within the plant cell. It is a polymer of glucose. Once the cellulose has been secreted by the cell and laid down in the cell wall, it is no longer involved in any of the chemical reactions of the cell. The cell wall remains intact even after the cell it surrounds dies. All that remains in dried wood, for example, are cell walls. The cells of the living tree which they once surrounded are all dead.

2-13 The structure of the cilium or flagellum as shown by (a) an electron micrograph of cilia in cross section and (b) a labeled diagram. (*Courtesy of John Kloetzel, Ph. D., University of Colorado.*)

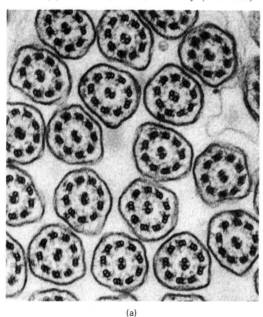

(a)

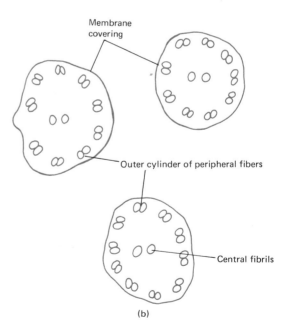

(b)

Membrane covering

Outer cylinder of peripheral fibers

Central fibrils

Cell Structure [45]

The Average Cell

Not all cells contain all the structures and organelles described in this chapter. Cells of higher organisms are highly specialized in order to perform particular functions for the organism. For study purposes, however, it is useful to construct a diagram of an "average" cell which is a composite of all the structures found in most cells (Fig. 2-14). A diagram of this sort allows not only a summary of the cellular structures discussed, but also a ready comparison of the relative sizes of all these bodies. The structures typical only of plant cells, the chloroplast and the cell wall, have not been included in this diagram.

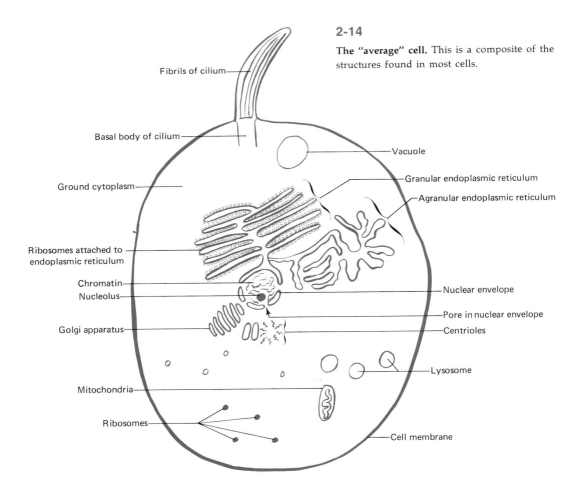

2-14

The "average" cell. This is a composite of the structures found in most cells.

The Origin of Life and the Cell

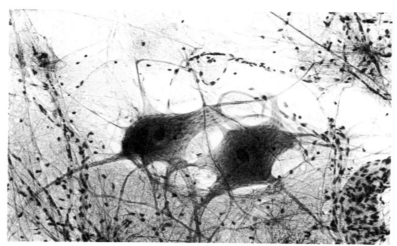

2-15

The general form of the nerve cell. It is specialized to its function in transmitting nerve impulses from location to location by the long branches which serve somewhat like wires in reaching over distances. (*Photomicrograph courtesy of Carolina Biological Supply Company.*)

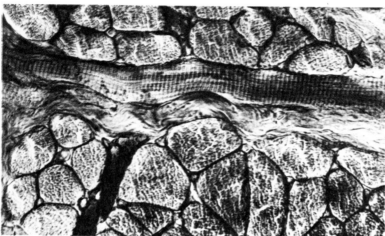

2-16

The general appearance of muscle. The wavy appearance comes from the ordered arrangement of the many protein fibers the cell uses to contract forcefully (note the longitudinal fiber across center of photomicrograph). (*Photomicrograph courtesy of Carolina Biological Supply Company.*)

2-17

The general form of bone cells. This cell is specialized to its function in secreting the bone matrix by a system of canaliculi that make it possible for the isolated cell to secrete the bone matrix without itself becoming isolated from the blood stream and other cells around it. (*Photomicrograph courtesy of Carolina Biological Supply Company.*)

In order to illustrate the extent that real cells deviate from this hypothetical construct of an "average" cell, three examples of specialized cells are presented: a nerve cell, a muscle cell, and a bone cell. The nerve cell has long branches which are useful to its function of "communication" (Fig. 2-15). The high content of protein fibers in a "striped" arrangement give the muscle cell the ability to contract with a great force (Fig. 2-16). The canaliculi which connect bone cells allow these cells to be distributed throughout nonliving bony material and still maintain contact with their necessary food and gas sources (Fig. 2-17).

Suggested Readings

FAWCETT, D. W. *An Atlas of Fine Structure, The Cell, Its Organelles and Inclusions,* Saunders, Philadelphia, Pa., 1966.

JELLINK, P. H. *The Cellular Role of Macromolecules,* Scott, Foresman, and Company, Glenview, Ill., 1965.

The Living Cell. Readings from *Scientific American,* W. H. Freeman and Company, San Francisco, Calif., 1965.

Chemical Composition of Cytoplasm 3

A knowledge of the chemical composition of cytoplasm is important to a meaningful understanding of the structure and function of the cell. Chemicals are classified as organic and inorganic compounds; cytoplasm is composed of both. Because compounds composed of a carbon skeleton are common to all living systems, they are called organic compounds. Inorganic chemicals are those without this carbon backbone. In spite of the nomenclature, organic compounds comprise only 10–30% of the weight of an average cell. The bulk of the cytoplasm is water which contains 1% dissolved inorganic compounds.

Inorganic Chemicals in Cytoplasm

Chemical evolutionists theorize that life began in the ocean. This is indeed plausible since the cell is 70–90% water which contains salts in ratios similar to those in seawater. The chemical reactions typical of living systems today occur very efficiently in this "seawaterlike" medium. It is reasonable to believe that these, or similar reactions and their reactants, occurred in the ocean and evolved into a cell.
Water

Upon dissolving in water, ionic compounds break apart into their component ions. Since ions are charged particles, the resulting solution conducts electricity. Because of this property, these ionic compounds are referred to as electrolytes. Electrolytes are classified
Electrolytes

by the kinds of ions they form in water. Acids produce hydrogen (H⁺) ions in solution, bases produce the hydroxyl (OH⁻) ion, and salts produce any ions but OH⁻ or H⁺. All three types, acids, bases, and salts, are found in the cell. Although electrolytes compose only about 1% of the cell weight, there are some 30 or more different electrolytes common to cytoplasm.

In spite of the low electrolyte content of the cell, electrolytes are very important to cell function. The concentrations of the ions formed are critical to the chemical function of the more complex organic compounds necessary for life. For instance, the activity of enzymes is very sensitive to the acid/base and salt concentrations of the cytoplasm.

The important ions in cytoplasm include sodium (Na^+), potassium (K^+), magnesium (Mg^{2+}), calcium (Ca^{2+}), chloride (Cl^-), bicarbonate (HCO_3^-), and phosphate (PO_4^{3-}). K^+, Na^+, Ca^{2+}, and Cl^- are present in a certain balance within the cell. This balance must be maintained for the cell to function properly. This again suggests that life might have evolved in the ocean. If, as postulated, chemical reactions of the cell evolved in the ocean, these reactions should be most efficient in the electrolyte concentration of this primordial sea; this is precisely what is observed.

Magnesium ion is vital to the cell for several reasons. It is the central ion in the chlorophyll molecules in plant cells. Without this ion the chlorophyll molecule does not absorb the solar energy the plant needs to live. The ribosomes of all cells also require Mg^{2+}. Ribosomes are important in protein synthesis.

Because of the chemical reactions shown in Fig. 3-1, the bicarbonate ion is important in the maintenance of the acid/base balance within the cell. If the cell becomes too acid, HCO_3^- and H^+ react to form H_2CO_3 and ultimately CO_2 and H_2O. Subsequent loss of CO_2 to the atmosphere successfully disposes of the excess acid (see Fig. 3-1, reaction 1). If the cell becomes too basic, this excess OH^- can be absorbed by the H_2CO_3, and the HCO_3^- is reformed (reaction 2).

The PO_4^{3-} ion is important to the energy exchange system of a cell. During the synthesis of adenosine triphosphate from PO_4^{3-}

3-1
The role of HCO_3^-/H_2CO_3 in regulating the acid and base content of the cell.

Reaction 1 HCO_3^- + H^+ ⟶ H_2CO_3 ⟶ H_2O + CO_2
              Bicarbonate   Acid

Reaction 2 OH^- + H_2CO_3 ⟶ H_2O + HCO_3^-
              Base   Bicarbonate

and adenosine diphosphate, energy is trapped and stored for later use within the cell.

Many other electrolytes, present in very small concentrations, are important to various cell functions. Some of the reactions of the cell can occur only when the organic molecules of the cell associate with these ions.

Organic Constituents of Cytoplasm

Organic compounds are composed of a skeleton of carbon atoms covalently bonded to one another. If the only other atoms bonded to this carbon backbone are hydrogen atoms, the compound is a hydrocarbon. Hydrocarbons are chemically unreactive. Other atoms (for example, O, N, or sulfur, S) or groups of atoms (for example, —COOH, —NH$_2$, —OH, or —SH) also bond to these carbon skeletons. Because of their chemical reactivity, these groups of atoms are referred to as functional groups. In chemical shorthand the inert or unreactive part of an organic molecule is denoted R—. It is connected to the functional groups where the chemical reactions occur. A list of the functional groups common to organic chemicals in the cytoplasm and examples of such cellular constituents is displayed in Fig. 3-2.

Although organic compounds are basically covalent compounds, some functional groups can ionize in water making these organic compounds electrolytes. As shown in Fig. 3-3 carboxyl groups, —COOH, can ionize to form H$^+$ ions in water and the amine groups, —NH$_2$, form OH$^-$ ions. Therefore, a compound possessing a —COOH group is an organic acid. An amine is an organic base.

Organic acids combine with other organic compounds to form larger molecules by losing water. These reactions are shown in Fig. 3-4. An acid and an amine combine to form an amide and water (reaction 1). Similarly, an acid and an alcohol react to form an ester and water (reaction 2). The resulting amide and ester do not ionize in water as the starting acids did. Many of the organic compounds in the cytoplasm, such as proteins, fats and carbohydrates, are formed by such dehydration syntheses.

Proteins

Although proteins comprise only 7–15% of the weight of the cytoplasm, they are very important to cell function and structure. All enzymes are proteins, and enzymes are compounds which enhance

3-2

Functional groups found in organic constituents of cytoplasm.

Functional group	Compound	Example from cytoplasm
—COOH Carboxyl	R—COOH Acid	$CH_3(CH_2)_{16}COOH$ Stearic acid

—NH$_2$
Amino
	R—NH$_2$
	Amine
		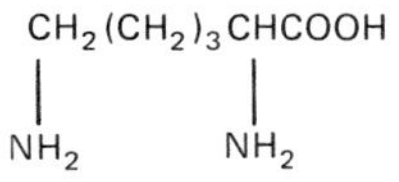 $CH_2(CH_2)_3CHCOOH$
		  | |
		  NH_2 NH_2
		Lysine

—OH Hydroxyl	R—OH Alcohol	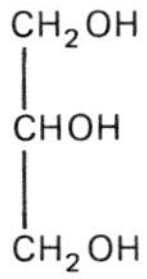 CH_2OH | $CHOH$ | CH_2OH Glycerine
—SH Sulfhydryl	R—SH Mercaptan	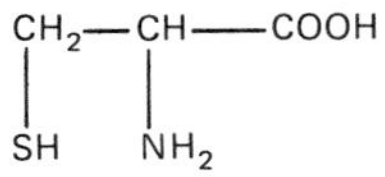 CH_2—CH——$COOH$ | | SH NH_2 Cysteine
—OPO$_3$H$_2$ Phosphoryl	R—OPO$_3$H$_2$ Phosphate ester	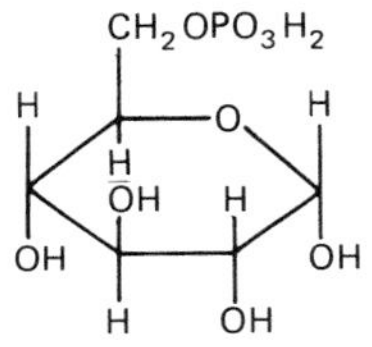 Glucose-6-phosphate

3-3

Organic electrolytes.

A carboxyl compound is an organic acid:

$$R—COOH \rightleftharpoons R—COO^- + H^+$$

An amine as an organic base:

$$R—NH_2 + H_2O \rightleftharpoons R—NH_3^+ + OH^-$$

3-4

Dehydration syntheses involving organic acids.

Reaction 1 $R—COOH$ + $R'—NH_2$ $\rightleftharpoons$ $R—\overset{O}{\overset{\|}{C}}—NHR'$ + H_2O
                 Acid   Amine   Amide

Reaction 2 $R—COOH$ + $R'—OH$ $\rightleftharpoons$ $R—\overset{O}{\overset{\|}{C}}—OR'$ + H_2O
                 Acid   Alcohol   Ester

the efficiency of the chemical processes within the cell. Besides possessing enzymatic activity, proteins are components of many of the vital cell structures such as mitochondria, chromosomes, chloroplasts, and cell membranes.

Proteins are long chains of amino acids strung together by amide linkages. An amino acid is an organic compound with both an amine group and a carboxyl group bonded to the same carbon atom:

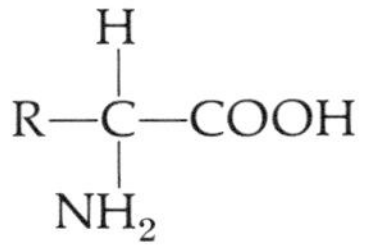

$$R-\underset{NH_2}{\overset{H}{C}}-COOH$$

The rest of the molecule can vary. There are about 20 different amino acids found in living organisms. Since an amino acid has both a carboxyl and an amine group in the same molecule, it can participate in reactions typical of acids as well as amines. Therefore, each end can form an amide linkage with another amino acid. As shown in Fig. 3-5 the carboxyl group of one amino acid (AA_1) can

3-5 **The formation of a protein from amino acids.** (a) The amino acid building blocks of a protein. (b) The reaction of an amino group of one amino acid with a carboxyl group of another to form an amide linkage and a molecule of water. An amide linkage between two amino acids is called a *peptide* linkage. (c) The resulting tripeptide.

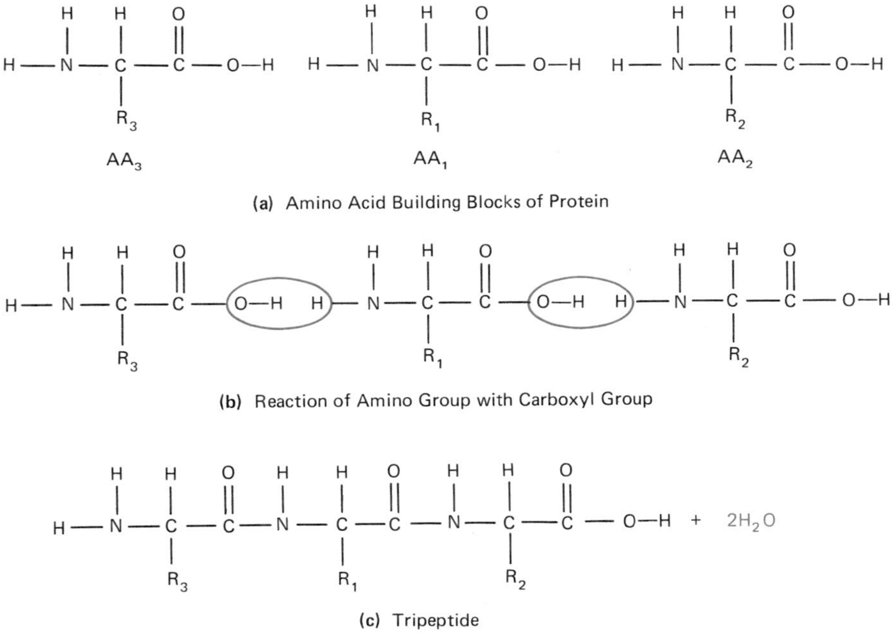

(a) Amino Acid Building Blocks of Protein

(b) Reaction of Amino Group with Carboxyl Group

(c) Tripeptide

Chemical Composition of Cytoplasm

form an amide with the amine of AA_2 and the amine of AA_1 can form an amide with the carboxyl of AA_3 (Fig. 3-5b). The amide linkages in proteins are called peptide bonds. In this way a polymer is built up with a carboxyl group on one end and an amine on the other. Since there are 20 different amino acids which are incorporated into proteins, the amino acid content and sequence can vary a great deal from protein to protein.

Fats Another important organic constituent of cytoplasm is fat. Cells store food materials in the form of droplets of fat, and some cell

3-6

The formation of fat from glycerol and fatty acids. (a) Glycerine plus three fatty acids. (b) The reaction of three fatty acids with glycerol to form a triester and three molecules of water. (c) The resulting fat.

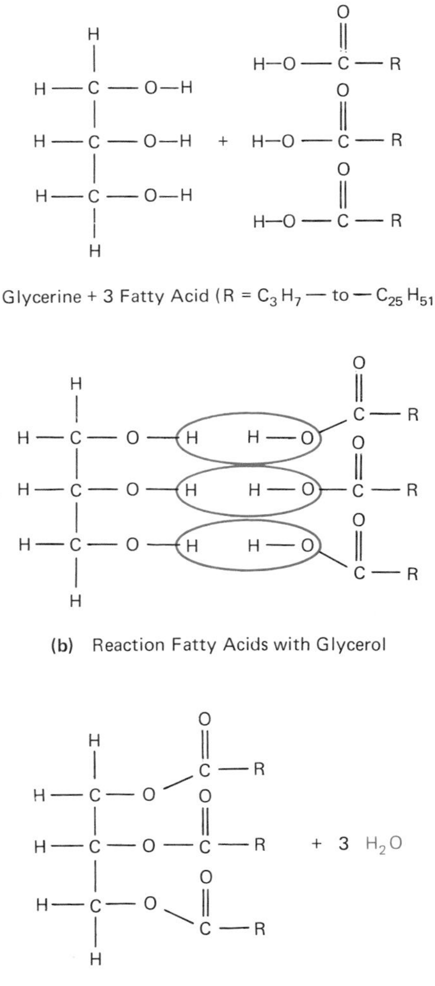

(a) Glycerine + 3 Fatty Acid ($R = C_3H_7 -$ to $- C_{25}H_{51} -$)

(b) Reaction Fatty Acids with Glycerol

(c) Fat

[54] The Origin of Life and the Cell

structures such as cell membranes contain fats. Fats are a group of compounds regarded as greasy or oily. They are more soluble in hydrocarbon and organic solvents like benzene, acetone, and chloroform than they are in water.

Fats are composed of fatty acids and glycerine. Glycerine is a chain of three carbon atoms. Bonded to each carbon atom of glycerine is a hydroxyl group; it is therefore a trialcohol (Fig. 3-6a). If each —OH group of glycerine reacts with an organic acid, a triester is formed along with three molecules of water (Fig. 3-6b). A fat is a triester of glycerine and three fatty acids (Fig. 3-6c). A fatty acid is composed of a long hydrocarbon chain bonded to a carboxyl group. There are many different fatty acids found in living systems. Fatty acids vary considerably in the number of carbon and hydrogen atoms in the hydrocarbon chain (from C_3 to C_{25}).

The final triester is mostly hydrocarbon in nature. It does not contain any functional groups which can ionize in water, so the fat is insoluble in water. A typical fat, beef fat, has a chemical formula of $C_{57}H_{110}O_6$ indicating the high hydrocarbon content.

Carbohydrates

Sugars are carbon chains (usually five or six carbon atoms long) with a C:H:O ratio of 1:2:1. Almost every carbon atom of any sugar is bonded to a hydrogen atom and an —OH group (Fig. 3-7, top). These chains can take a cyclic or ring form as shown at the bottom in Fig. 3-7. Carbohydrate is a class of cellular constituents which includes sugars and chains of sugars of varying lengths. Carbohydrates are important to the cell structurally and energetically. For instance, cellulose of the plant cell wall is a carbohydrate. Sugars are the chemical energy source for most cells. The breakdown of sugar gives off energy which the cell uses to drive other cell reactions. Plants can make sugars, a process which requires energy, by trapping energy from sunlight in a process called photosynthesis.

The sugar chains of carbohydrates differ in sugar components and chain length. For example, glucose is a single six-carbon sugar (C_6). Maltose and sucrose are dimers composed of two C_6 sugars. Maltose is composed of two identical molecules of glucose, but sucrose (cane sugar) is made up of glucose and fructose. Starch is a polymer composed of many molecules of glucose. One molecule of water is lost each time a glucose molecule links onto a starch chain (Fig. 3-8). This process can continue until a large starch chain emerges. Some starch chains develop side chains which are similar to, but shorter than, the main chain.

3-7

The forms of typical sugar. Glucose and fructose can assume a cyclic or ring formation. Glucose forms a ring of six atoms, a pyranose. Fructose forms a ring of five atoms, a furanose.

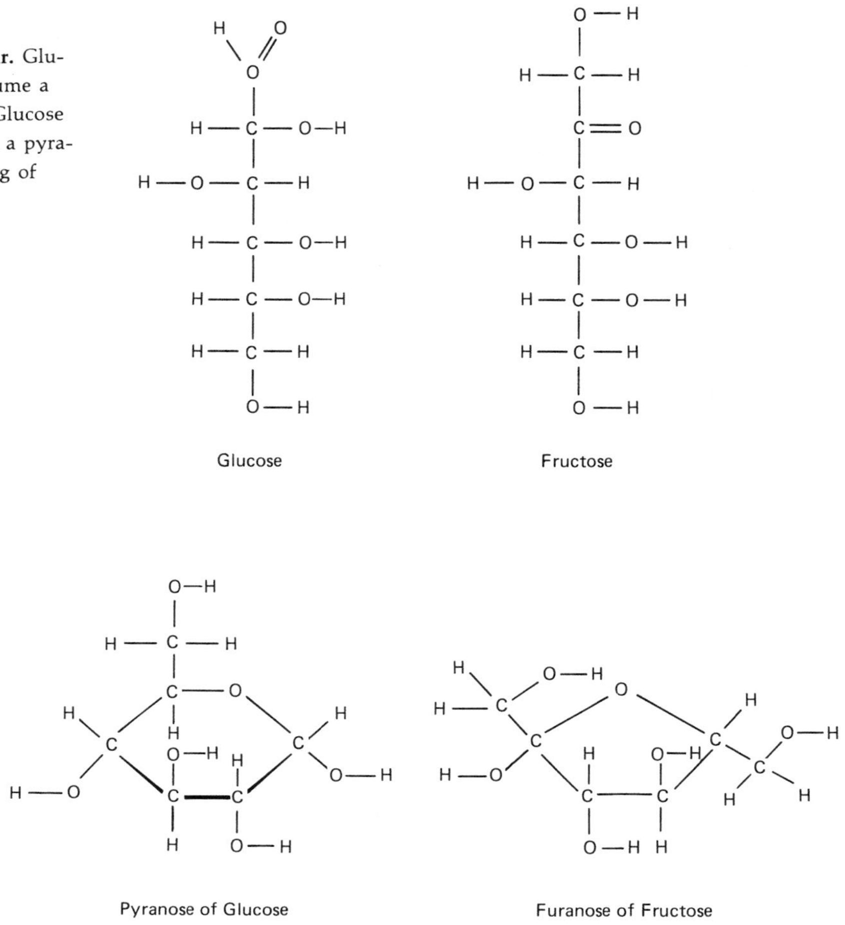

Glucose

Fructose

Pyranose of Glucose

Furanose of Fructose

Adenosine Phosphates

Adenosine phosphates are important constituents of the cytoplasm because they are the energy transfer agents within the cell. They are composed of three kinds of compounds: an organic base, adenine; a sugar, ribose; and one to three inorganic phosphate groups. The adenine and the phosphate are bounded to the first and the fifth carbon atoms, respectively, of the C_5 sugar. This combination is depicted in Fig. 3-9. If the adenosine phosphate has only one phosphate, it is adenosine monophosphate (AMP). Adenosine diphosphate (ADP) and triphosphate (ATP) have two and three phosphates, respectively.

It takes energy to form the diphosphate from AMP and inorganic phosphate. This energy is released when the diphosphate reverts to AMP and PO_4^{3-}. The same is true for the ADP + $PO_4^{3-} \rightleftharpoons$ ATP conversion. The last two phosphate linkages of ATP are referred to as "energy-rich" bonds. (This is denoted by a "squiggly" bond,

(a) Glucose Molecule Building Blocks of Starch

(b) Water is Lost

(c) Starch Molecule

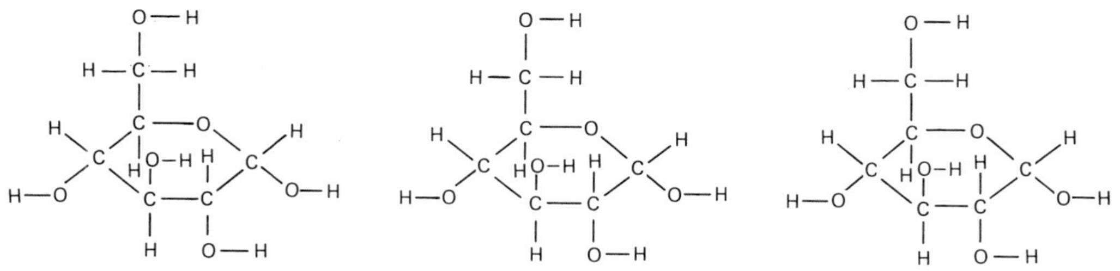

3-8 **The formation of starch from units of glucose.** (a) The glucose molecule building blocks of starch. (b) Water is lost as one molecule of glucose reacts with another. (c) The resulting starch molecule.

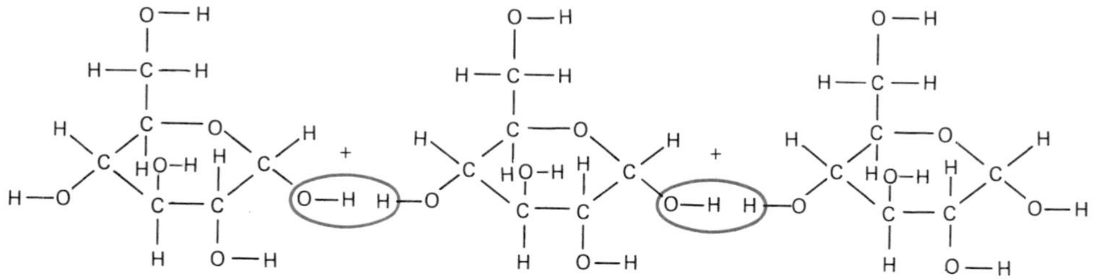

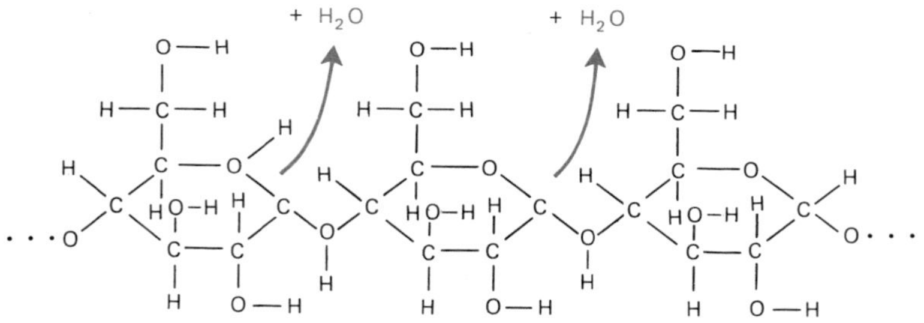

3-9 **The structure of adenosine triphosphate (ATP).**

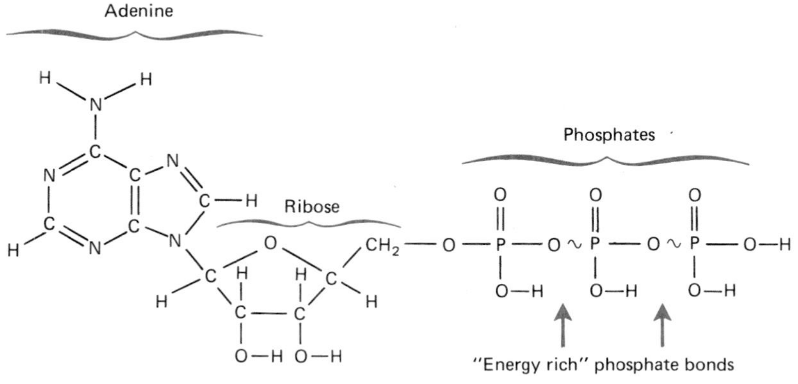

~PO_4.) When energy is given off by a reaction occurring in the cell, it can be "stored" and not lost to the surroundings if it is used to form ATP from ADP and PO_4^{3-}. As will be seen in Chapter 7, 38 molecules of ATP are formed during the breakdown of one molecule of glucose, the cell's chemical fuel. This stored energy can then be released when the cell needs it if the ATP reverts to ADP or AMP and inorganic PO_4^{3-}.

Nucleic Acids Nucleic acids are the most complex macromolecules in the cell. Like the other macromolecules, proteins and carbohydrates, nucleic acids are polymers. However, the polymerized unit in the nucleic acids is itself a composite molecule called a nucleotide. A nucleotide is a compound composed of an organic base and a phosphate group bonded to a five-carbon sugar. The AMP described in the last section is a nucleotide.

There are two kinds of nucleic acids in cells—deoxyribonucleic acid, DNA, and ribonucleic acid, RNA. These differ mostly in the sugar component and slightly in the base composition. The nucleotides which comprise DNA are composed of the five-carbon sugar deoxyribose bonded to a phosphate at C_5 and to one of the four nitrogenous bases, thymine, adenine, cytosine, or guanine at C_1. RNA nucleotides are made up of the sugar ribose with a phosphate bonded at C_5 and one of the four bases, uracil, adenine, cytosine, or guanine bonded at C_1. These two sugars and five bases are pictured in Fig. 3-10. The bases adenine and guanine have a similar backbone composed of two fused rings. These bases are purines. Cytosine, thymine, and uracil have only a single-ring structure. These smaller bases are pyrimidines.

The nucleotides are strung together into the nucleic acids by a dehydration reaction between the phosphate group of one nucleotide and the hydroxy group at C_3 of the sugar of another nucleotide (Fig. 3-11). In this manner long chains of nucleotides are built up. Nucleic acid molecules can reach a total length of several millimeters. The bases (B) hang out from this sugar (S)–phosphate (P) backbone.

$$-S-P-S-P-S-P-S-P-S-$$
$$\;\;\;|\quad\;\;\;|\quad\;\;\;|\quad\;\;\;|\quad\;\;\;|$$
$$\;\;\;B\quad\;B\quad\;B\quad\;B\quad\;B$$

In this structure the base sequence and content can vary considerably. Chemical analysis of the base content of DNA from various organisms indicated that the base content was not totally variable.

Although the different bases occurred in different ratios in DNAs from different organisms, the molecular amount of adenine (A, a purine) always equalled that of thymine (T, a pyrimidine), and the molecular amount of guanine (G, a purine) equalled that of cytosine (C, a pyrimidine). This structure does not account for this relationship. X-ray crystallographic analysis indicates that the DNA molecule actually consists of two such strands coiled together (Fig. 3-12).

In this coil or "double helix" the two strands are lined up opposite one another so that the -S-P-S-P- "backbones" are on the outside of the helix and the bases are in the center of the helix facing one another. That is, a purine base of one strand is paired to a pyrimidine base on the other strand. More specifically, every A in one

3-10 The building blocks of the nucleotides of the nucleic acids of the cell.

Sugars

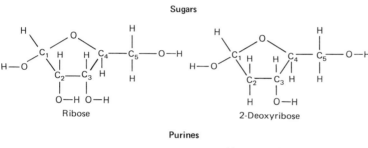

Purines

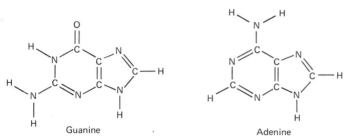

Pyrimidines

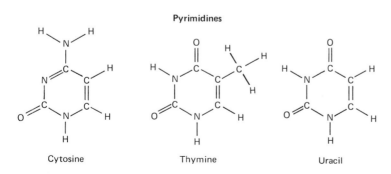

Chemical Composition of Cytoplasm [59]

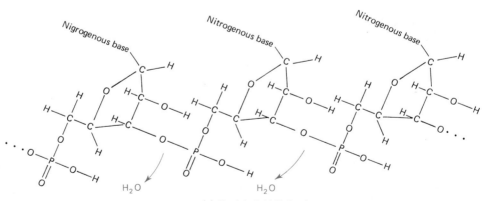

3-11

The formation of a nucleic acid from nucleotides by dehydration synthesis. (a) The nucleotide building blocks of nucleic acids. (b) The removal of —OH and —H to form H_2O and bond the nucleotides together into a nucleic acid. (c) The nucleic acid molecule formed by the dehydration synthesis of many nucleotides.

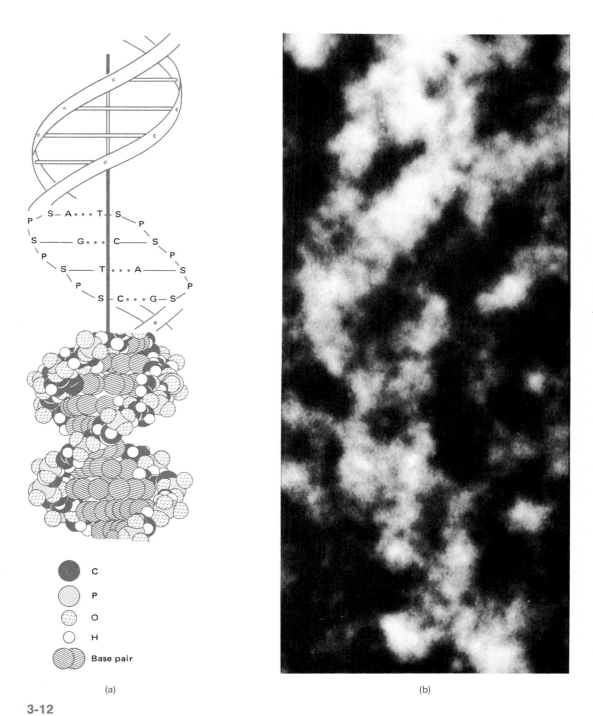

3-12

The double helical form of the deoxyribonucleic acid (DNA) molecule as seen from (a) a diagram and (b) and electron micrograph and special preparation of the molecule. The diagram gives three different ways in which the molecular arrangement of DNA is often given.

[61]

3-13
The hydrogen bonding in water.

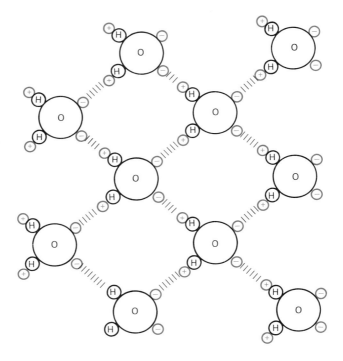

H represents hydrogen atom of water molecule
O represents oxygen atom of water molecule
⊕ represents positive polarity charge
⊖ represents negative polarity charge
|||| represents hydrogen bonding

strand is paired to a T in the other. Similarly every C in one strand is opposite a G in the other.

$$\begin{array}{cccccccccc}
-P-S-P-S-P-S-P-S-P-S- \\
||||| \\
AGTCG \\
\\
TCAGC \\
||||| \\
-P-S-P-S-P-S-P-S-P-S-
\end{array}$$

This structure explains why the molecular amount of A equals that of T and the molecular amount of C equals that of G. This particular pairing also explains the uniform distance found between the two strands. Although the purine bases are larger than the pyrimidine bases, each pair includes a purine and a pyrimidine so that the two pairs are about the same size.

Apparently this double-stranded structure of DNA is held together because these paired bases are weakly bonded to one another. In a covalent bond an electron pair is "shared" between the two atoms of the bond. If the atoms are identical as in a C-C bond, this "sharing" is literal. If the atoms are different, however, this electron pair will be attracted more to one atom than the other, since the nuclei of the different atoms have different electronic configurations. Oxygen and nitrogen attract an electron pair more than either carbon or hydrogen does in the O-C, N-C or O-H, N-H bonds. Such imbalances are described as though partial charges existed. In the O-C, N-C, O-H, and N-H bonds the oxygen and nitrogen are slightly negative and the carbon and hydrogen slightly positive. The slightly negative oxygen and nitrogen atoms in molecules containing these bonds will attract the slightly positive hydrogen atoms from other molecules containing such bonds. This weak interaction is called hydrogen bonding. The hydrogen bonding of water is illustrated in Fig. 3-13. The partial charges in the nitrogenous bases are such that hydrogen bonding occurs only between A and T and between C and G (Fig. 3-14).

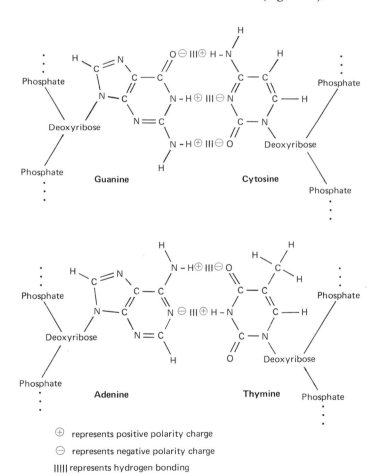

3-14

Hydrogen bonding of the nitrogenous bases of the nucleotides in DNA. Note that the configuration of guanine fits that of cytosine, and adenine fits that of thymine. This accounts for the specific pairing pattern of the nucleotides in DNA.

Chemical Composition of Cytoplasm [63]

Elemental Composition of Cytoplasm

Approximately 98% of the weight of cytoplasm is composed of only four elements, carbon, hydrogen, oxygen and nitrogen. A close look at the elemental composition of the main molecular constituents of the cell—water, proteins, fats, and carbohydrates—indicates why this might be so.

- ☐ Water is composed only of hydrogen and oxygen.
- ☐ Protein is made up primarily of carbon, hydrogen, oxygen and nitrogen.
- ☐ Fats and carbohydrates contain only carbon, hydrogen, and oxygen.

Adenosine nucleotides and nucleic acids contain phosphorous in the form of phosphate. Both of these compounds are vital to cell function, the former for energy transfer and the latter for protein synthesis, yet the phosphorous content is less than 1% of the cell weight. There are other elements vital to cell function that are present only in minute amounts. These elements are called trace elements.

It is theorized that carbon, hydrogen, nitrogen, and oxygen were the abundant and reactive chemicals at the surface of the earth mass from which life developed. This was the environment that led to the chemicals and chemical reactions now associated with living systems. It is logical that organisms be composed mainly of these four elements.

Suggested Readings

BENNETT, T. P., and EARL FRIEDEN, *Modern Topics in Biochemistry*, Macmillan, New York, 1966.

DEBUSK, A. GIB. *Molecular Genetics*, Macmillan, New York, 1968.

GOLDSBY, R. A. *Cells and Energy*, Macmillan, New York, 1967.

Part II

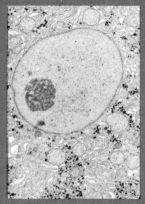

Cell Function

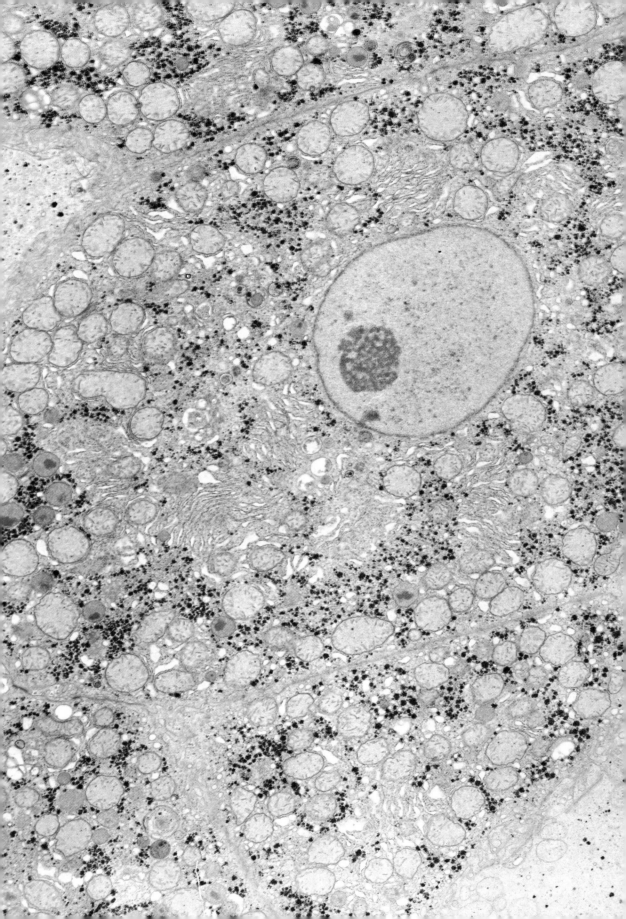

The Role of DNA in Cell Function 4

Deoxyribonucleic acid, DNA, is the hereditary material of the cell. That is, DNA is the cellular component which not only determines the traits an organism may exhibit but which also is exactly duplicated during reproduction so that the offspring organisms will exhibit the basic characteristics of the parental species. On the chemical level an organism's characteristics are due to the chemical reactions occurring within its cell(s). DNA must, then, govern the reactions or functions of the cell by some chemical mechanism.

The mechanism by which DNA controls the cell is to determine what proteins are made. Most reactions of a cell only occur if there is a specific enzyme present which will accelerate that reaction. All enzymes are proteins and each one can only enhance the rate of a certain cellular reaction. It is believed that this specificity of function is due to the three-dimensional shape of each enzyme. Because of its shape, only certain reactant molecules can "fit" into the reactive center of each enzyme. The three-dimensional shape of a protein is, in turn, due to its chemical structure; that is, which amino acids it contains and how they are arranged in the polymeric chain. DNA apparently directs the specific ordering of amino acids into proteins.

The structure of DNA was discussed in detail in Chapter 3. It is the intent of this chapter and the next to show how DNA behaves at the molecular level to account for its role in the transmission and expression of hereditary information. The structure of the DNA molecule gives it the chemical stability, the ability to be replicated, and the ability to carry information which allows DNA to function as the hereditary component of cells.

Electronmicrograph of cross section of whole cell. (*Courtesy of A. E. Vatter, University of Colorado, Webb Waring Institute.*)

Relative Stability of DNA

DNA is a relatively stable molecule. The chemical information that it carries therefore remains intact from generation to generation. This means that offspring organisms have the same basic characteristics as the parental species. DNA is, however, not so stable that it cannot change at all. Both chemical and physical forces can cause the nitrogen bases of DNA to change. Any change in the DNA structure alters the chemical information the DNA carries.

The changes in DNA structure brought about chemical and biological evolution. As the structure of DNA changed, the functions it could direct within a system also changed. Any chemical aggregate (chemical evolution) or organism (biological evolution) which could maintain itself efficiently in its environment could outlast other systems which were not as efficient in that environment. The DNA from any system that could not last was lost. The DNA from "successful" systems, however, was not. The small changes in DNA structure, then, allowed for the appearance of a wide variety of organisms, while the essential stability of DNA structure insured the continued appearance of the most successful organisms from generation to generation.

Replication of DNA

The double helix structure of DNA allows for an exact duplication of DNA by template replication. The sugar and phosphate units in the backbones of the two nucleotide strands of DNA are bonded together quite strongly. These bonds do not break very easily. The hydrogen bonds between the purine and pyrimidine bases, however, are much weaker. These weaker bonds are the ones which hold the two strands of the double helix together. When DNA replicates, the strands of the double helix slowly unwind and separate as the hydrogen bonds between complementary base pairs break (Fig. 4-1a).

These freshly exposed bases are then free to hydrogen bond with triphosphodeoxyribonucleotides present in the nucleus. The triphosphodeoxyribonucleotide contains a three-phosphate chain attached to the deoxyribose unit of the nucleotide at the same location where the single phosphate is normally attached. The three-phosphate chain is important as a source of energy for the subsequent bonding of the sugar and phosphate groups to form a chain. Because adenine (A) can only hydrogen bond to thymine (T) and guanine (G) can only hydrogen bond to cytosine (C), any base in either of the DNA strands will only hydrogen bond to a nucleotide containing the same base that it was bonded to in the double-stranded form (Fig. 4-1b). As the two strands of a DNA molecule continue to unwind, the exposed bases can continue to hydrogen bond specifically to such complementary nucleotides. In the presence of the enzyme DNA-polymerase, the triphosphonucleotides which are hydrogen bonded to bases that are adjacent on the DNA strand bind together between the sugar and phosphate groups to form a -phosphate-sugar-phosphate- backbone (Fig. 4-1c). In this way each strand of the original DNA molecule acts as a template for the formation of a new complementary strand. Two DNA molecules, each identical to the original molecule, result (Fig. 4-1d).

The Coding System of DNA

DNA is only found in the nucleus, yet it directs the protein synthesis which occurs in the cytoplasm outside the nucleus. The structure of DNA must therefore carry specific information in some kind of a chemical "code" which can be transcribed and transferred to the site of protein synthesis. The only variability in the structure of DNA is in the base sequence along the nucleotide chains. This sequence of bases, then, must somehow carry the chemical code which dictates the amino acid sequence in proteins.

A great amount of recent research effort has been directed at determining how this information is coded and how the cell transcribes it so that it reaches the cytoplasm. The experimental evidence indicates that the information in DNA is transferred to RNA during the synthesis in RNA in the nucleus. It is the RNA that carries this information to the cytoplasm. Presumably the code is the same for both nucleic acids, since they both have the same basic nucleotide chain structure.

The Code Unit

The code unit of DNA and RNA is the nucleotide. In DNA it is the deoxyribonucleotide, and in RNA it is the ribonucleotide. For both DNA and RNA there are only four kinds of nucleotides, each differing only in the nitrogen base. The four bases in DNA nucleo-

4-1a The hydrogen bonds between complementary base pairs in DNA break.

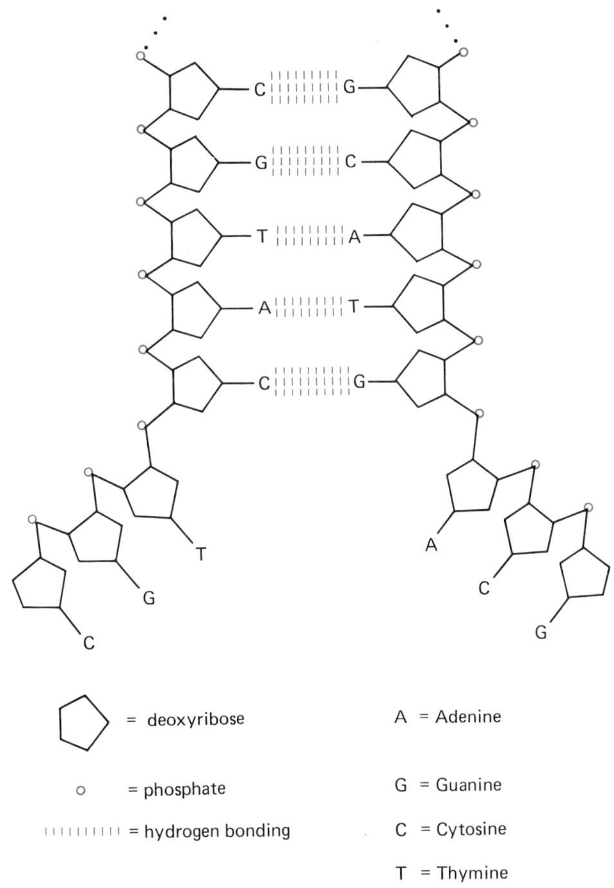

4-1b Triphosphodeoxyribonucleotides hydrogen bond to the freshly exposed bases on unwinding DNA stand.

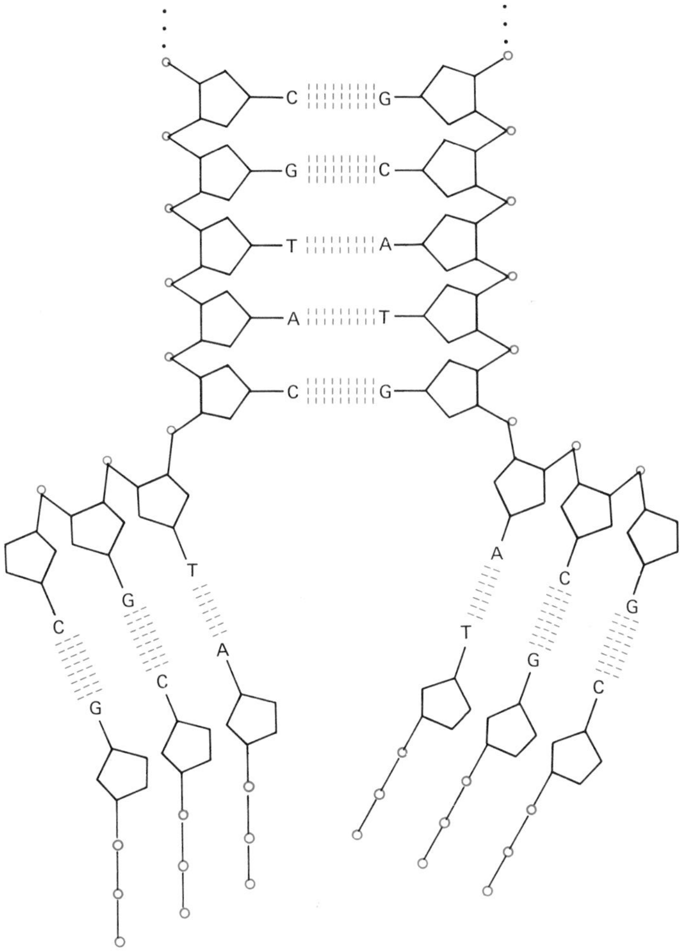

A = Adenine
G = Guanine
C = Cytosine
T = Thymine

⬠ = deoxyribose
○ = phosphate
||||||||| = hydrogen bonding

[70]

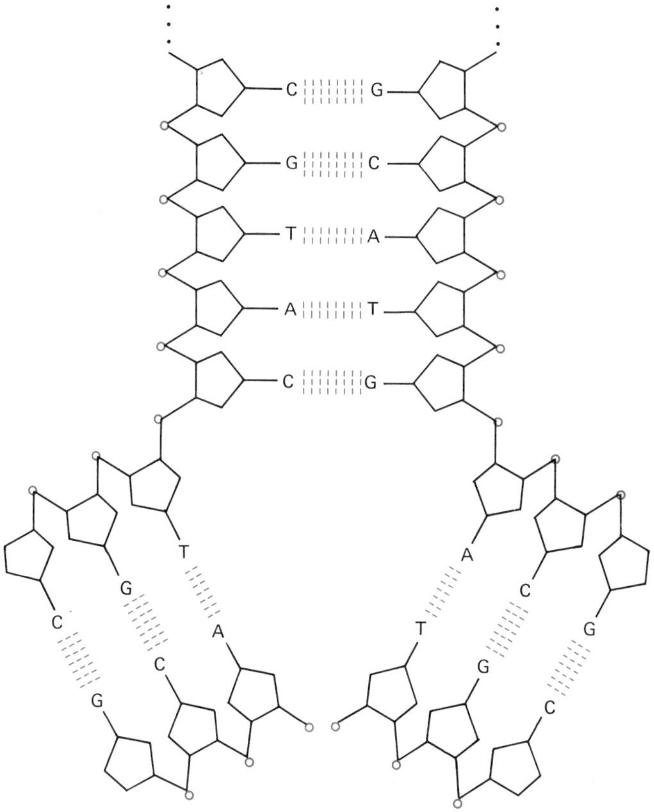

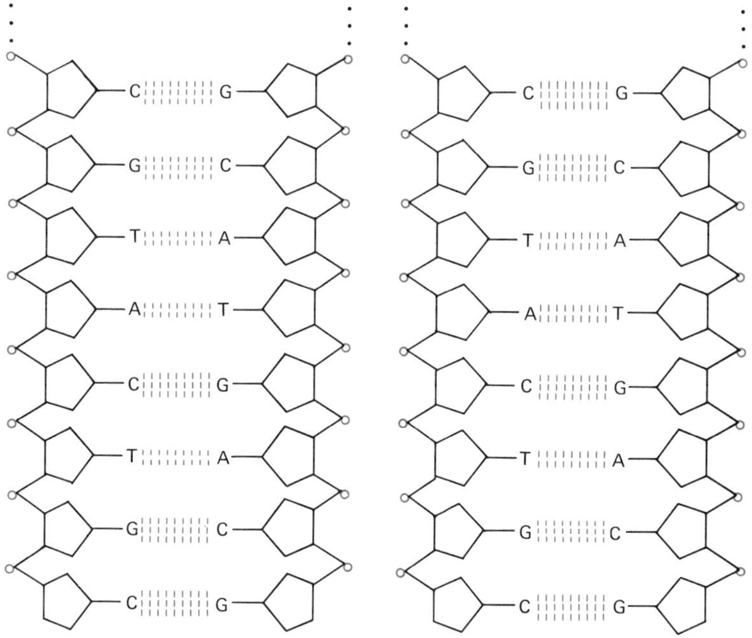

4-1d As a result of this process, when the two strands of the original DNA molecule are completely separated, two new molecules of DNA result. Each molecule is identical to the original molecule.

4-1c A sugar-phosphate-sugar backbone forms by interaction of the adjacent triphosphodeoxyribonucleotides in the presence of the enzyme DNA polymerase.

[71]

tides are adenine, thymine, guanine, and cytosine. Those in RNA nucleotides are adenine, uracil, guanine, and cytosine. The proteins in most organisms, however, are composed of some 20 different amino acids. If nucleic acids carry information for all 20 amino acids, the coding system must depend on combinations of these nucleotides. This is analogous to the International Morse Code in which two code units, the dot and the dash, are combined in 26 different ways to represent the 26 letters of the alphabet.

The Coding System

The coding system of the nucleic acids must lie in the sequential combinations of nucleotides along the nucleic acid chains. The coding system is assumed to be the same for both DNA and RNA, since the code for RNA is derived from DNA and their chemical structures are basically alike. All subsequent examples will be given in terms of RNA (that is, with base sequences composed of A, U, C, and G rather than those of A, T, C, and G) since this code actually directs protein synthesis within the cytoplasm.

It is presumed that the code is only "read" in one direction, so that the sequence of bases along the nucleotide chain is important; that is, CU is different than UC. Linear combinations of two nucleotides can only give 16 sequentially different combinations. Each nucleotide, when followed by any one of the four nucleotides gives four sequentially different combinations. In all there are 16 such combinations:

$$A + \begin{cases} A = AA \\ U = AU \\ G = AG \\ C = AC \end{cases} \quad U + \begin{cases} A = UA \\ U = UU \\ G = UG \\ C = UC \end{cases} \quad G + \begin{cases} A = GA \\ U = GU \\ G = GG \\ C = GC \end{cases} \quad C + \begin{cases} A = CA \\ U = CU \\ G = CG \\ C = CC \end{cases}$$

Since there are 20 amino acids to be coded for incorporation into proteins, there are not enough doublet combinations to code all 20 amino acids unless the code is ambiguous. An ambiguous code is one in which a single code unit codes more than one thing. In this case some nucleotide doublets would have to code more than one amino acid. An ambiguous code would certainly lead to variations in amino acid sequence. For instance, if one doublet, D_1, coded two different amino acids, valine and glycine, another doublet, D_2, coded lysine and leucine, and a third, D_3, coded serine and arginine, the following amino acid sequences could result from a $D_1D_2D_3$ code:

Valine-Lysine-Serine
Valine-Lysine-Arginine

Valine-Leucine-Serine
Valine-Leucine-Arginine
Glycine-Lysine-Serine
Glycine-Lysine-Arginine
Glycine-Leucine-Serine
Glycine-Leucine-Arginine

That is, eight different amino acid sequences could result from the same coded segment of nucleic acid. Since such variation in protein structure is not found, it is believed that the coding system is not ambiguous. The doublet, therefore, is not the basic code combination.

There are 64 sequentially different combinations of nucleotide triplets. These can be generated by adding each nucleotide to the 16 doublets: $4 \times 16 = 64$. Clearly, then, there are enough combinations of triplets to code the 20 amino acids unambiguously. Experimental evidence which is beyond the scope of this book indicates that the code combination of the nucleic acids is the triplet of nucleotides. This triplet of nucleotides is called a codon. Since there are more possible codons than amino acids to code, the code can be, and has been found to be, degenerate. That is, several codons may code for the same amino acid. Experimentally, 61 codons have been found to code amino acids, and almost all amino acids are coded by at least two different codons (Fig. 4-2).

A degenerate code can direct the synthesis of a specific amino acid sequence. For example, suppose codons CUG, CUA, UUG, and UUA all code the amino acid leucine. Similarly, the codons UUC or UUU code phenyl alanine, and GGU, GGC, GGA, and GGG code glycine. The following combinations of codons would all code the same amino acid sequence of leucine-phenylalanine-glycine:

CUG-UUC-GGU	CUG-UUU-GGU	CUA-UUC-GGU
CUG-UUC-GGG	CUG-UUU-GGG	CUA-UUC-GGG
CUG-UUC-GGA	CUG-UUU-GGA	CUA-UUC-GGA
CUG-UUC-GGC	CUG-UUU-GGC	CUA-UUC-GGC
CUA-UUU-GGU	UUG-UUC-GGU	UUG-UUU-GGU
CUA-UUU-GGG	UUG-UUC-GGG	UUG-UUU-GGG
CUA-UUU-GGA	UUG-UUC-GGA	UUG-UUU-GGA
CUA-UUU-GGC	UUG-UUC-GGC	UUG-UUU-GGC
UUA-UUC-GGU	UUA-UUU-GGU	
UUA-UUC-GGG	UUA-UUU-GGG	
UUA-UUC-GGA	UUA-UUU-GGA	
UUA-UUC-GGC	UUA-UUU-GGC	

	Second base of triplet				
	U	C	A	G	
U	UUU } Phe UUC UUA } Leu UUG	UCU UCC } Ser UCA UCG	UAU } Tyr UAC UAA UAG	UGU } Cys UGC UGA UGG Tryp	U C A G
C	CUU CUC } Leu CUA CUG	CCU CCC } Pro CCA CCG	CAU } His CAC CAA } Glun CAG	CGU CGC } Arg CGA CGG	U C A G
A	AUU AUC } Ileu AUA AUG Met	ACU ACC } Thr ACA ACG	AAU } Aspn AAC AAA } Lys AAG	AGU } Ser AGC AGA } Arg AGG	U C A G
G	GUU GUC } Val GUA GUG	GCU GCC } Ala GCA GCG	GAU } Asp GAC GAA } Glu GAG	GGU GGC } Gly GGA GGG	U C A G

First base of triplet — Third base of triplet

Amino Acids
Glycine (Gly)
Serine (Ser)
Glutamine (Glun)
Arginine (Arg)
Alanine (Ala)

Threonine (Thr)
Phenylalanine (Phe)
Tyrosine (Tyr)
Valine (Val)
Aspartic acid (Asp)

Cysteine (Cys)
Histidine (His)
Leucine (Leu)
Asparagine (Aspn)
Methionine (Met)

Tryptophan (Try)
Isoleucine (Ileu)
Glutamic acid (Glu)
Lysine (Lys)
Proline (Pro)

4-2 The genetic code. This table lists the 64 possible triplet combinations of the four bases (A, G, U, C) and the amino acids which 61 of these triplets "code."

There are 32 different codon series that will code the same specific amino acid sequence. What is important is that a specific amino acid sequence can be synthesized with a degenerate code.

Since there are 64 possible codons and only 20 possible amino acids to be coded, there may be codons which do not code amino acids. Apparently there are three such codons. These are called nonsense codons. These codons have no use in determining the amino acid sequence in the synthesis of proteins. They may serve, however, to mark a break between proteins. In such a function these codons would serve as "punctuation marks" in the "reading" of the genetic information.

There exists the possibility that the triplet code is overlapping. That is, one nucleotide could serve in more than one codon as is illustrated on the following page:

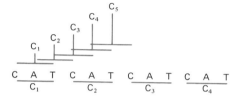

If the code is overlapping, the codon sequence in this example "reads" CAT, ATC, TCA, CAT, ATC, TCA, and so forth. If the codons are discrete units, sequentially arranged, the code reads instead CAT, CAT, CAT, CAT, and so forth. Experimental evidence indicates that the code is not overlapping and that the triplet codons are discrete units, each following the next.

The evidence piling up in favor of the sequential code of discrete triplet codons lined up next to one another indicates that there are no single nucleotides which act as "commas" or markers between codons. If a nucleotide of the triplet served as a comma, the remaining doublets could not code all the amino acids as discussed earlier.

The triplet coding structure thus far described for nucleic acids is apparently universal for all organisms. The genetic code, that is, the specific amino acid which each codon codes, is essentially universal for all organisms as well. This was discovered from the observation that a cell-free system of *E. coli* bacteria, if given RNA from several different organisms, produces the particular protein that the foreign RNA codes.

The Translation of the Coding System

There are two intermediary steps by which the base sequence of DNA gets translated into the amino acid sequence in proteins. First, the base sequence of DNA gets transcribed into the base sequence of RNA during the synthesis of RNA in the nucleus. Then, the coded information in RNA directs the incorporation of amino acids into proteins in the cytoplasm. The mechanism of RNA synthesis will be described here, but that of protein synthesis will be discussed in the next chapter.

The process of synthesizing DNA which requires DNA-polymerase, triphosophodeoxyribonucleotides, and existing DNA is called *template replication*. The synthesis of RNA which requires the enzyme RNA-polymerase, triphosophoribonucleotides, and existing DNA is called *transcription*. In the process of transcription, a single strand of DNA serves as template for complementary base pairing with the triphosphoribonucleotides in much the same way as it does in DNA replication with the triphosphodeoxyribonucleo-

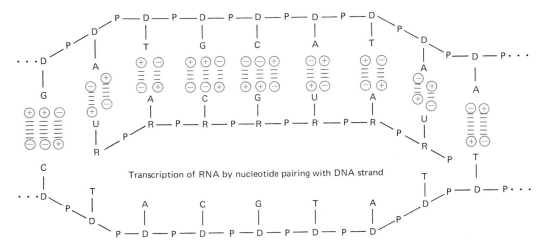

4-3 A schematic diagram of the synthesis of RNA. The bases of ribonucleotides pair to the unpaired bases of one strand of DNA. A ribose-phosphate-ribose backbone forms when ribonucleotides which are hydrogen bonded to adjacent bases on the DNA strand bond to one another. This process is similar to that of DNA replication except that only a small segment of DNA uncouples, only one strand is transcribed, and the pairing of nucleotides are ribonucleotides.

tides. It is believed, however, that in the synthesis of RNA only a short segment of the DNA unwinds, and only one strand of the partially unwound DNA serves as template for the ribonucleotides. The guanine of the DNA strand hydrogen bonds to a cytosine ribonucleotide; the thymine of the DNA strand pairs with an adenine ribonucleotide; and the cytosine and adenine of DNA strands pair with guanine and uracil ribonucleotides, respectively. Ribonucleotides hydrogen bonded to adjacent bases on the DNA template strand get bonded together between the ribose unit of one and the phosphate unit of the other. In this way the -ribose-phosphate-ribose- backbone of the RNA chain is synthesized. This synthesis requires the presence of the enzyme RNA-polymerase. The process results in the formation of RNA with an accurate transfer of the genetic information of the DNA (Fig. 4-3).

The strand of RNA synthesized in this way is usually much shorter than that of the DNA it was transcribing. Therefore the RNA strand carries only a fraction of the information of the DNA molecule. RNA molecules in the cell are usually less than 1 μ long, but DNA molecules are sometimes as long as 2 mm (2000 times as long). Apparently RNA molecules carry information necessary for one or, at best a few, particular functions, whereas DNA carries information for many different functions.

Suggested Readings

BENNETT, T. P., and EARL FRIEDEN. *Modern Topics in Biochemistry,* Macmillan, New York, 1966.

DEBUSK, A. GIB. *Molecular Genetics,* Macmillan, New York, 1968.

5 Protein Synthesis

The synthesis of specific proteins within the cell is essential to both the function and structure of the cell. It is the chemical information contained in the nucleotide sequence of DNA which directs this synthesis of specific proteins. This information must be relayed from the nucleus to the site of the protein synthesis in the cytoplasm and utilized there in a very precise manner. Furthermore, all proteins needed for cell function and structure during the cell's lifetime are not always present in the cell nor are they all being synthesized there at the same time. In general, specific proteins are only synthesized and present in the cell when they are required for a specific function or structure. Evidently cells possess a control mechanism which dictates when specific proteins are to be synthesized.

A great deal of recent experimental activity has been directed toward answering the questions of how and when protein synthesis occurs. The mechanism, control, and importance of protein synthesis within the cell which this experimentation has indicated will be described in this chapter.

The Mechanism of Protein Synthesis

Protein synthesis takes place on the ribosomes in the cytoplasm. Ribosomes are very small, compact masses of RNA and protein. They are shaped in the form of two connected spherical particles. The RNA of a ribosome, rRNA, is important to protein synthesis, but its mechanism of action is not known as yet.

As discussed in Chapter 4, the chemical information in the nuclear DNA is transcribed to a single strand of RNA when RNA is synthesized using a part of a DNA strand as template. This RNA molecule, which carries the codons that specify the amino acid sequence for a specific protein, is called messenger-RNA, mRNA. It moves from the nucleus into the cytoplasm and one end of it gets attached to a ribosome. This mRNA-ribosome complex acts as a template for the formation of proteins.

A third type of RNA, transfer-RNA, tRNA, is the actual decoder. It provides the actual link between the nucleotide codon and the amino acid it codes. The tRNA molecule is much smaller than either mRNA or rRNA. On the average, tRNA is a chain of only about 80–90 nucleotides.

Because it is small, the three-dimensional structure of tRNA could be determined. It has been found to be a single strand which has doubled back on itself in any of several possible ways (Fig. 5-1).

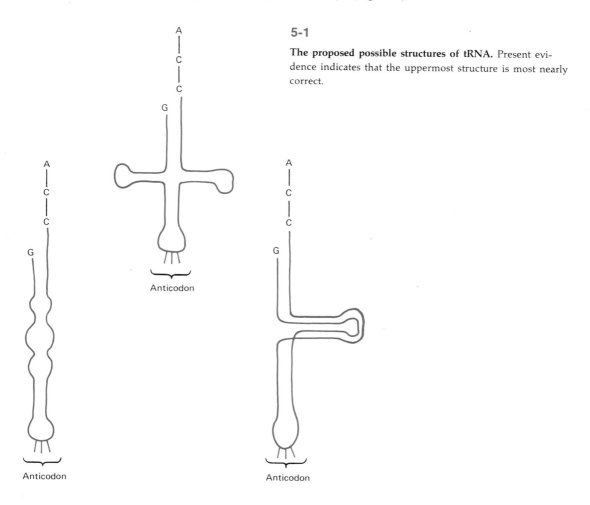

5-1

The proposed possible structures of tRNA. Present evidence indicates that the uppermost structure is most nearly correct.

Protein Synthesis [79]

5-2

The first step in protein synthesis. An amino acid must be activated before it can interact with tRNA.

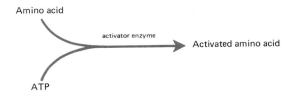

The overall configuration, is essentially a double-stranded molecule with a loop at one end and the two ends of the RNA strand at the other. The forces holding the doubled-back strands together are again hydrogen bonds between the bases on the two halves of the molecule. The end with the loop in it necessarily has bases which are not paired. This end is called the anticodon, since three unpaired bases there can pair to those of a complementary mRNA codon. At the other end of the tRNA molecule are the two strand ends. In all tRNA molecules, one strand end terminates with a G nucleotide and the other with a CCA nucleotide sequence.

Any amino acid which gets incorporated into a protein must apparently first be activated by reacting with an energy-rich ATP molecule in the presence of a specific enzyme (Fig. 5-2). Once the amino acid is activated, it can attach itself to the adenine base at the end of the tRNA molecule which terminates in the CCA sequence (Fig. 5-3). This step is apparently very specific, but the mechanism of this specificity is not known. A particular amino acid will only attach itself to tRNA molecules with certain anticodons.

5-3

This diagram illustrates how an activated amino acid gets attached to a tRNA molecule. By some mechanism the anticodon of the tRNA molecule determines which activated amino acid gets attached.

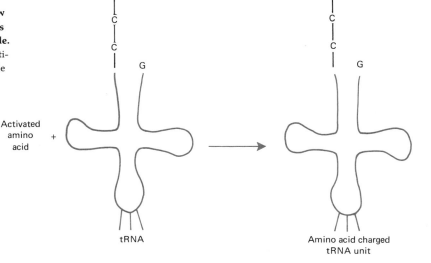

Cell Function

Specifically, the tRNA molecules to which an amino acid attaches are only those which possess the anticodons to the mRNA codons which "code" that amino acid. The amino acids and the codons which "code" them are displayed in Fig. 4-2.

The point of contact between mRNA and the ribosome serves as the point of tRNA binding to the mRNA. The tRNA-amino acid unit is moved to this tRNA binding site on the ribosome-mRNA complex. The tRNA-amino acid unit that binds there is that one with the anticodon which has the complementary base sequence for the mRNA codon at the site. For instance, if the mRNA codon at the binding site were AAG, only tRNA molecules with the anticodon UUC would bind there. The amino acid at the other end of the tRNA molecule gets added onto the end of the protein chain which is growing on the ribosome at that end of the binding site. Once this amino acid transfer has occurred, the uncharged tRNA unit moves away. The ribosome moves along the mRNA strand to the next codon. The process is repeated, as another tRNA-amino acid, with an anticodon complementary to this next codon, binds to the mRNA at the binding site. This amino acid is added onto the protein, and so forth (Fig. 5-4).

Since a specific amino acid is always attached to a tRNA with a particular anticodon, that amino acid will always be added to a growing protein when the complementary mRNA codon appears at the tRNA binding site. In this way a specific amino acid sequence is built up because of the sequential array of codons in the mRNA.

Several ribosomes can move along an mRNA molecule at the same time, forming several points of contact for different complementary tRNA-amino acid units to bind simultaneously. This unit is called a polysome. Several identical proteins can then be assembled at once in this manner (Fig. 5-5). The amino acid chain attached to the ribosome which has progressed the farthest down the mRNA chain will be the longest at any particular time in this process.

The Importance of Proteins to the Cell

Proteins are found to some degree in all the major structures of the cell. They occur in the membranes of various cell structures, such as the cell membrane, the Golgi apparatus, the endoplasmic reticulum, the mitochrondria, the lysosomes, the chloroplasts, and the nuclear envelope. Proteins also compose a part of ribosomes, nucleoli, centrioles, spindle fibers, and chromosomes. Although

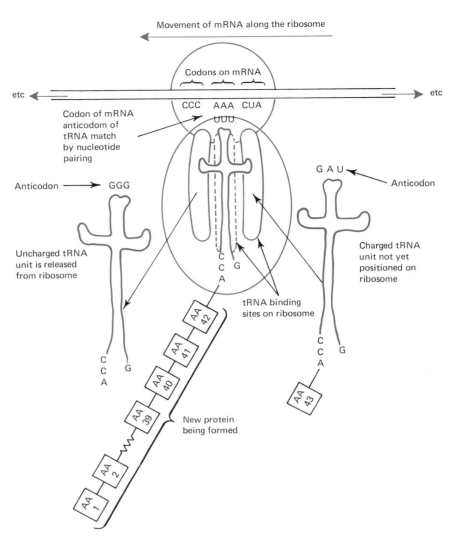

5-4 This diagram illustrates how the mRNA-ribosome complex directs the assembly of a specific sequence of amino acids into a protein chain. The tRNA-amino acid unit with an anticodon complementary to the codon on the mRNA at the binding site positions itself at the binding site. The amino acid at the other end of the tRNA molecule is added to the chain of amino acid residues that is being formed at the end of the binding site. The uncharged tRNA unit is then released from the binding site and moves into the cytoplasm. The ribosome then moves to the next codon on the mRNA and another tRNA-amino acid unit with the appropriate complementary anticodon binds to that codon. Another amino acid is added to the protein, and so forth.

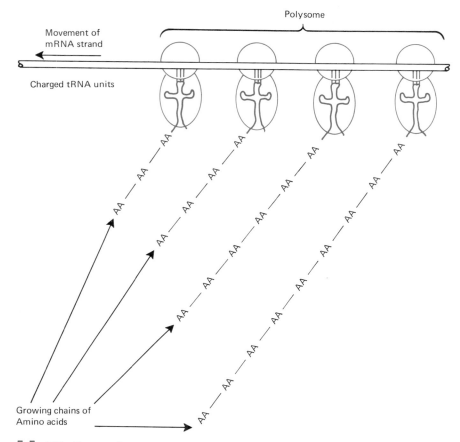

5-5 This diagram illustrates the course of protein synthesis when several ribosomes are attached to the same mRNA molecule at the same time. Several identical proteins can be synthesized at once by this polysome.

these structures are not totally protein in content, cellular structure would be totally disrupted without protein.

Proteins are vitally important to cell function as well as to cell structure, as all enzymes are composed mostly of protein. Enzymes enhance the rates of specific chemical reactions so that they can occur at the temperatures and reactant concentrations found in the cells. Most of the reactions of the cell would not occur at a significant rate under cellular conditions without enzymes to speed them up. An enzyme, then, is a protein catalyst, as it is a protein which accelerates a reaction without being altered itself.

The mechanism by which most enzymes are believed to act is to bring the reactant sites into contact with one another by the formation of some sort of an enzyme-reactant complex. Either during or after the reaction, this complex breaks apart to regenerate the unchanged enzyme and yield the reaction products. Since enzymes are unchanged in catalysis, they can function repeatedly. Therefore, only a small amount of enzyme may be needed by the cell to accelerate a particular reaction.

Presumably it is the structure of an enzyme that determines which reactants can bond chemically (usually weakly) with the enzyme to form the enzyme-reactant complex. The three-dimensional structure is determined by the sequence of amino acids along the protein chain of the enzyme. The long polypeptide chain of each enzyme bends back around on itself in a specific way. The interactions between the amino acid groups hanging off the polypeptide chain cause this specific interwinding. Some amino acid groups which are adjacent spatially (but not sequentially along the chain) can bond together strongly with covalent bonds if they contain reactive groups. Some form hydrogen bonds, which, of course, are weaker. These interactions, both weak and strong, maintain a specific three-dimensional structure.

It is because of this specific three-dimensional structure that enzymes are so specific in the chemical reactions which they catalyze. Presumably only a certain reactant can "fit" at the active site, the position of the enzyme-reactant complex formation. In effect, the active site is custom made for a particular reactant molecule. This relationship of the enzyme and reactant is analogous to that of a lock and key (Fig. 5-6).

Many of the properties of enzymes are related to this three-dimensional structure and its maintenance. The hydrogen bonds which maintain the three-dimensional structure are easily ruptured with heat. Because of this, many enzymes are permanently deactivated by heat; that is, after heating they no longer possess catalytic activity. The acidity of the medium can also affect the activity of an enzyme. Many of the amino acid groups along a protein chain change structure between acid and basic conditions. These changes can easily alter the chemical nature of the active site and/or the three-dimensional structure. For instance, the enzyme in saliva which decomposes starch can only catalyze this reaction at near neutral conditions. It is deactivated by the acids of the stomach.

Some enzymes are only able to catalyze a reaction in the presence of another substance. This substance is called a cofactor. The same substance can serve as cofactor for several different enzymes which catalyze related reactions. One group of cofactors which are important to cellular activity are the electron carriers. Many of these are dinucleotides, consisting of an adenine nucleotide and another

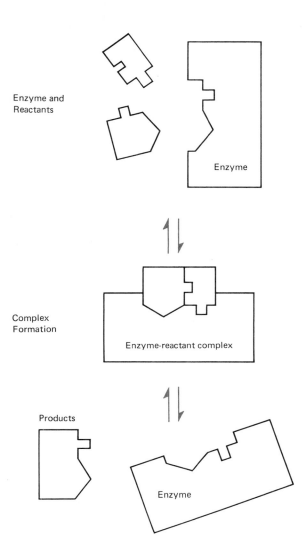

5-6

A schematic representation of the theory of enzyme specificity. An enzyme can catalyze only a certain reaction because its three-dimensional structure can only accommodate certain reactants. Once it brings these reactants together, they can interact to form products.

nucleotide. The second nucleotide possesses a functional group which is capable of accepting or donating electrons under certain cellular conditions. Common cofactors of this type, which are particularly important in photosynthesis and respiration, are NAD (nicotinamide adenine dinucleotide), NADP (nicotinamide adenine dinucleotide phosphate), and FAD (flavin adenine dinucleotide). The complete chemical structure of NAD is presented in Fig. 5-7a only to illustrate the dinucleotide structure of these compounds. Only the functional groups which actually take part in electron transfer are displayed for all these cofactors in order to illustrate their mechanisms of action (Fig. 5-7b).

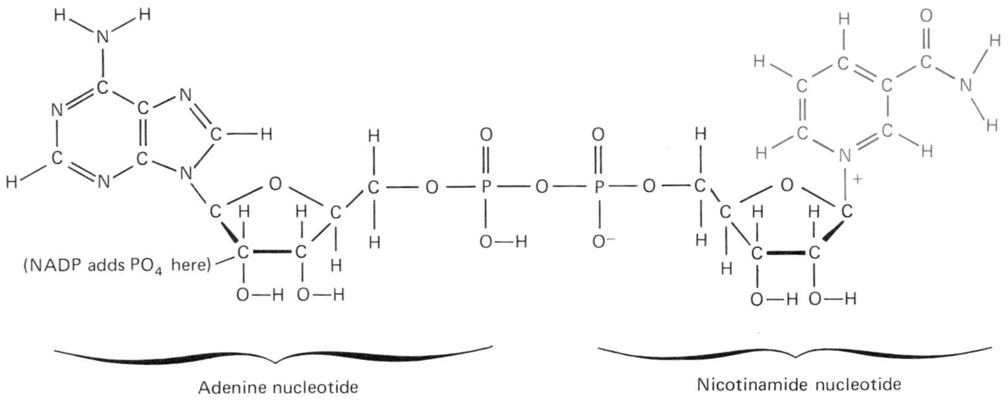

5-7 (a) The structure of NAD (nicotinamide adenine dinucleotide). NADP (nicotinamide adenine dinucleotide phosphate) has the same basic structure, but has a phosphate group in the adenine nucleotide as indicated. (b) The electron transfer functions of the electron carrier cofactors. In these structures R— represents the rest of the dinucleotide which is not involved in electron transport.

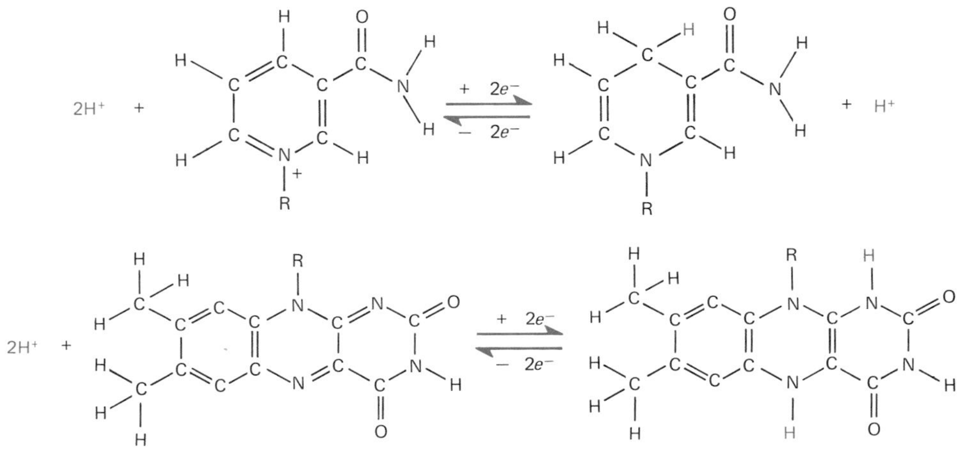

It is mainly due to the nature and activity of enzymes that cells behave as they do. It is because of enzymic catalysis that cells can both carry on reactions at rates which require higher temperatures outside the cell and carry on a complex sequence of chemical reactions. The sensitivity of cells to both temperature and acid-base balance is due to the sensitivity of enzyme function to these variables.

A few examples of the effects on the human body caused by enzymatic deficiencies should illustrate the importance of enzyme function to the cell. Individuals with phenylketonuria do not have the capability of making the enzyme phenylalanine hydroxylase. This enzyme utilizes a common amino acid, phenylalanine, which is present in many proteins and is therefore ingested with normal food consumption. Phenylalanine builds up in the blood of any individual who cannot make the enzyme to decompose it. The build-up of phenylalanine causes brain damage and feeble mindedness for this individual. The albino condition is also caused by an enzyme deficiency. An albino individual cannot form the enzyme which transforms phenylalanine into the pigment, melanine. Without this pigment, the individual has no coloration. Galactosemia is a condition in which the individual cannot synthesize enzymes which utilize galactose, a sugar in milk. The individual with this deficiency suffers malnutrition.

Since enzymes are so important to cell function, protein synthesis is very important to the organism. The mechanism of protein synthesis shows how dependent cell function is on the hereditary material, DNA. The DNA apparently also controls protein synthesis so that a cell only synthesizes the amount and kinds of enzymes it needs at that particular time.

Control of Protein Synthesis

A gene is a section of nucleic acid which carries the coded information in its codons to govern the formation of one protein. Genes apparently fall into natural units called operons. An operon is composed of adjacent genes (on the same DNA molecule) which control and direct the formation of proteins that often regulate related processes. Each gene of an operon which directs the formation of a particular protein is a structural gene of the operon. The gene which helps control the function of these structural genes is called the operator gene (Fig. 5-8). It is apparently through the action of the operator gene on the operon that a cell produces only the kind and amount of proteins that it needs to function at any one time. Although the cell has the capability to produce more enzymes, it does not because of this operator gene control.

An operator gene cannot control protein synthesis by itself, a regulator gene is also needed. The regulator gene need not be adjacent or even near-by the operon it helps to control. One regula-

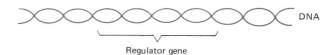

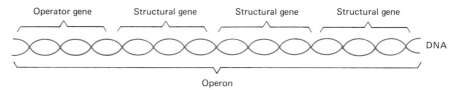

5-8 **The location of the genes of an operon and the regulator gene for the operon.** The structural genes of an operon are adjacent to one another along the DNA molecule. The operator gene is adjacent to the structural genes of the operon. The regulator gene which codes the production of a repressor is located on a different DNA molecule.

tor gene can help control the action of several operons. A regulator gene causes the formation of a repressor substance which can in some way influence an operator gene to either start or stop the action of the operon which it controls. To do this, there must be some mechanism by which the repressor substance can change form according to the cell's needs. Thus, when certain proteins are needed by the cell, the repressor substance is in a form which allows the operator gene to start the action of the operon which directs formation of these needed proteins. Similarly, when these proteins are no longer needed, the repressor substance must be in another form which acts with the operator gene to "turn off" the operon which causes their synthesis. There are two different repressor-operon systems by which the nature of the repressor substance reacts to the cell's needs: the inducible operon system and the repressible operon system.

An inducible operon system is one in which a particular protein-enzyme is only synthesized by the cell when a reactant in the reaction which that enzyme accelerates is present in the cell. A reactant in a reaction catalyzed by an enzyme is called a substrate for that enzyme. The inducible operon system gets its name, because a substrate induces, or causes, the formation of its own enzyme, that is the enzyme which enhances the cell's utilization of that substrate (Fig. 5-9).

In inducible operon systems, the repressor substance caused by the regulator gene acts on the operator gene to repress the operon function. This repressor, however, can be "deactivated" if the inducing substance bonds weakly to it. This induces the synthesis of the necessary enzyme. Once the enzyme is synthesized, the inducer will be used up by enzyme action. There will no longer be any inducer to bind to the repressor; the repressor will be reactivated, and the operon will turn off.

In the repressible operon system, the repressor substance cannot act on an operator gene alone to repress operon function. A corepressor is required to "activate" the repressor so that it can stop

5-9 **The theoretical balancing mechanism of the inducible operon system.**
When the substrate is present it inactivates the repressor and turns the operator "on" (b), whereas when the substrate is not present the repressor is active and the operator is turned "off" (a).

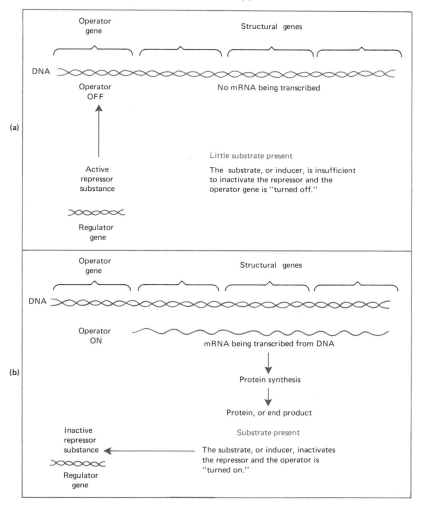

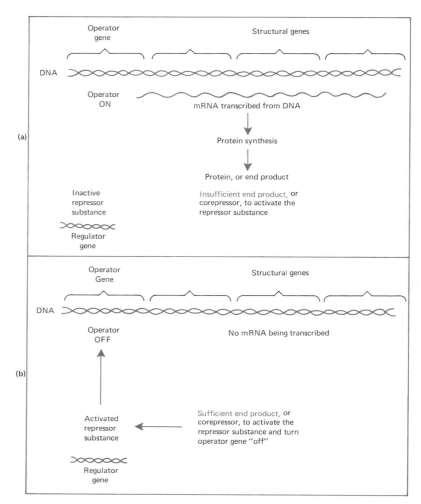

5-10 **The theoretical balancing mechanism of the repressible operon system.** When insufficient product, or corepressor, is present to activate the repressor substance, the operon is turned "on" (a); whereas when product, or corepressor, is in excess, the repressor substance is activated and the operon is turned "off" (b).

the action of the operon. In general, the corepressor is the end product of the reaction which the enzyme resulting from operon function catalyzes. When there is enough of a particular protein in a cell, the reaction it catalyzes will proceed to form the end product. This product bonds weakly to the repressor substance and activates it. The combination "turns off" the operon. If the protein formed by the operon is needed by the cell, no activated repressor would be available, and the operon would function. The site at which the activated repressor acts is, again, the operator gene (Fig. 5-10).

Both of these control mechanisms have been observed in bacteria and other lower organisms. There is some evidence that they are also operative in higher organisms such as flowering plants. It is assumed that they represent the general patterns for control of protein synthesis in most organisms. These mechanisms certainly allow for the cellular production of only the kinds and quantities of proteins that are needed at any one time.

6 Photosynthesis

Many of the steps along which life theoretically originated as described in Chapter 1 require energy. Energy was required both for the formation of the chemical compounds which aggregated into living systems and for the maintenance of these systems and their functions once they developed. The energy which was used to synthesize the chemicals which composed those primitive living systems apparently came from heat, lightning, and radiation. These energy sources, however, are and were not suitable for the maintenance of the chemical aggregates which were functionally so delicately balanced. This is certainly true for cells as they now exist. These energy sources supply so much energy that they are disruptive to cells. The cell, requires a constant supply of energy in rather small increments. The cell gets such an energy supply by controlling, with enzymes, the breakdown of a molecule containing chemical energy, a food-fuel. Sugars, starch, fats, and proteins are all such food-fuels. The energy stored within the bonds during the synthesis of these food-fuels is slowly released as each bond in these substances is broken, one by one.

The food-fuels of early life forms were those present in the rich organic mixture in which these forms arose. They were formed by chance synthesis brought about by the heat, lightning, or radiation of the early earth. Once a living system exhausted the foodstuff of its environment, it lost its capacity to maintain its function and itself. It is interesting to postulate how often this might have happened until a system developed which was able somehow to synthesize these foodstuffs within itself from readily available materials. This system would require a mechanism for trapping energy in order to drive such syntheses. The living systems that developed this capability were the ancestors of the green plants. The process they utilized to trap energy and synthesize foodstuffs to insure their own maintenance was photosynthesis.

Photosynthesis is the process by which green plants are able to utilize light energy from sunlight to synthesize organic food-fuels from the simple, readily available materials, H_2O and CO_2. The overall chemical equation for this transformation is

$$6\ CO_2 + 12\ H_2O \xrightarrow{light} C_6H_{12}O_6 + 6\ O_2 + 6\ H_2O$$

The $C_6H_{12}O_6$ is a simple sugar and is representative of the organic food-fuels produced. However, many other materials are also synthesized from this process. The organic compounds produced by green plants through photosynthesis are used as energy sources for the entire biologic community, both plant and animal.

Photosynthetic Requirements

Photosynthesis takes place within the chloroplasts of green plants. These organelles contain chlorophyll and many enzyme systems, some of which contain electron carriers. All these compounds are necessary for photosynthesis. In addition to the ingredients within the plant, photosynthesis requires CO_2, H_2O, and light from the environment.

Chlorophyll is a generic term for several substances with very similar molecular structures. All types of chlorophyll possess the same complex ring structure composed of a carbon-nitrogen backbone. In the center of this ring structure is a magnesium atom. The two most common types of chlorophyll, chlorophyll a and chlorophyll b, differ only slightly in the organic groups which are connected to this basic ring structure (Fig. 6-1). Chlorophyll is the pigment in plants responsible for their green color. It absorbs red light; hence, the light reflected by a plant looks green because it is "missing" the colors of light absorbed by the plant. It is the light energy that is absorbed by the chlorophyll that drives the synthesis of organic foodstuffs within plants.

A compound absorbs light of certain energy because that amount of energy can excite one of the electrons of that compound to a higher energy state. If the photosynthetic factory (that is, the enzyme and electron carrier systems) were not present within the chloroplast, the energy that the chlorophyll absorbs would either be lost from the chlorophyll as heat or reemitted as light as the

6-1

The molecular structure of chlorophyll a and chlorophyll b. The —CHO group replaces the —CH_3 group as shown here, for the form of chlorophyll b.

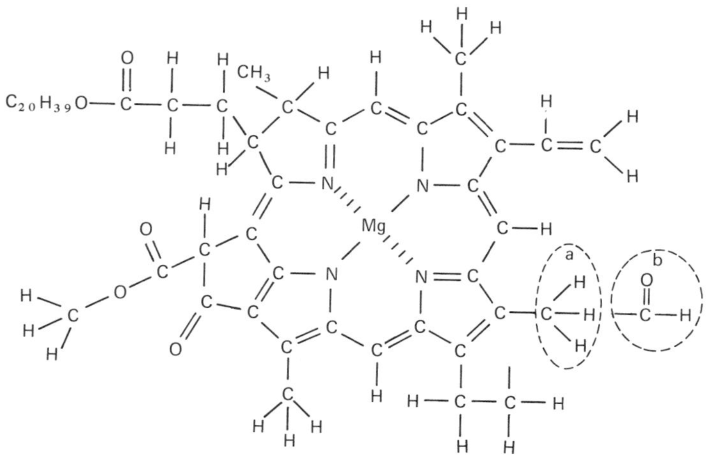

excited electron lost the extra energy and returned to its ground state. With the photosynthetic factory, however, this energy is not lost and, instead, is utilized to drive the formation of chemical bonds.

The energy is ultimately used to bind CO_2 and hydrogen from H_2O into energy-rich organic foodstuffs. This aspect of photosynthesis is called CO_2 fixation. The oxygen in H_2O gets liberated as gaseous O_2, a "waste" product of photosynthesis.

The Process of Photosynthesis

There are two basic processes of photosynthesis, the "light reactions" and the "dark reactions." The light reactions are those processes which are *directly* caused by the absorption of light energy by the chlorophyll. The chlorophyll ultimately returns to the unexcited state, but the energy it absorbed triggers the synthesis of the energy-rich compounds ATP and $NADPH_2$ with the aid of the enzyme–electron carrier systems. ($NADPH_2$ is the reduced form of NADP, reduction being the general chemical process of adding electrons, together with protons in this case.) These compounds are the chemical energy sources for the dark reactions. The dark reactions are those reactions which result *indirectly* from the light absorption, since they are energetically driven by the direct products of light absorption. Because of the production of ATP and $NADPH_2$ resulting from light absorption, CO_2 and hydrogen from H_2O get incorporated into organic foodstuffs. This CO_2 fixation and the processes which follow it constitute the dark reactions. Dark reactions can occur in both light *or* dark.

Light Reactions

The mechanisms of the light reactions are still very much in question. Apparently there are two different light reactions. Both of them involve electron transport. In one reaction the high-energy electron is ejected from the "excited" chlorophyll molecule and is transferred to a series of electron carriers. Electron carriers are substances which have a functional group in their molecular structure capable of both accepting and donating electrons (see Fig. 5-7b). As an electron is transferred from an electron carrier at one energy level to another at a lower energy level, the electron loses some of its energy. This energy causes the synthesis of an ATP molecule from ADP and inorganic phosphate, an energy-requiring reaction. Ultimately the electron loses all the extra energy obtained from the absorption of light by the chlorophyll and returns to the chlorophyll molecule in the ground, or unexcited, state. The chlorophyll molecule is then ready to be reexcited, and the whole process repeated. Because the system returns to its original state once the absorbed light energy is harnessed into ATP production, this reaction is called cyclic photophosphorylation (Fig. 6-2).

In the other light reaction there are two light-absorbing chlorophyll systems and two light-absorption steps (Fig. 6-3). As in the first reaction, both light-absorption steps result in the transfer of a high-energy electron from an "excited" chlorophyll system to an electron carrier system. However, unlike the first reaction, the ejected electron in both steps does not return to the chlorophyll system from which it was ejected. Instead there is a net transfer of electrons via an electron-transfer system from one chlorophyll

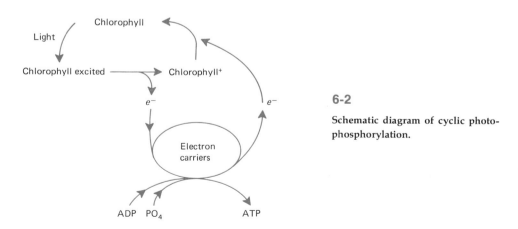

6-2

Schematic diagram of cyclic photophosphorylation.

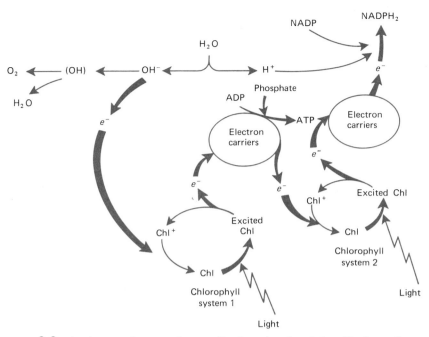

6-3 A schematic diagram of noncyclic photophosphorylation. The heavy line denotes electron flow.

system, composed mainly of chlorophyll b, to the other, composed mostly of chlorophyll a, and from the chlorophyll a system to an electron carrier, the cofactor NADP. This cofactor retains the electron by the addition of H^+ from H_2O to form $NADPH_2$. The ultimate source of electrons which restores the chlorophyll in the first system to its original state is the readily available ingredient, H_2O. The loss of an electron by the first chlorophyll system causes water to give an electron to this electron-deficient chlorophyll system. The mechanism for this reaction, called the photolysis of water is not known. This chlorophyll system can then absorb light again. The energy that is released during the transfer of an excited electron from the first chlorophyll system through an electron carrier system to the second chlorophyll system drives the synthesis of ATP from ADP and inorganic phosphate. Since neither ejected electron returns to its original source and there is instead a net electron transfer from H_2O to $NADPH_2$ which accompanies the formation of ATP, this process is called noncyclic photophosphorylation.

The proton produced by the photolysis of water is used in the $NADPH_2$ formation, along with the energetic electron. The (OH) byproducts from this photolysis rearrange to produce H_2O and gaseous O_2. The O_2 is a "waste" product of photosynthesis and is released by the plant into the atmosphere. The appearance in the atmosphere of this O_2 byproduct of photosynthesis significantly changed the course of chemical evolution from what it was in the originally oxygen-deficient environment. The changes which resulted are part of the "oxygen revolution."

The formation of both ATP and $NADPH_2$ is effected by the absorption of light energy by chlorophyll. These primary products of photosynthesis are used to power the subsequent incorporation of CO_2 into organic food-fuels in the process of CO_2 fixation.

Dark Reactions (CO_2 Fixation)

In CO_2 fixation, CO_2 reacts with a five-carbon sugar, C_5 which is energized with two phosphate groups (ribulose diphosphate). The C_6 molecule which forms is unstable and splits into two C_3 molecules of phosphoglyceric acid, PGA. The reaction of $NADPH_2$ and ATP with PGA forms phosphoglyceraldehyde, PGAld. Some of the PGAld gets used to build up more C_5 sugars to continue the CO_2 fixation cycle. The rest gets incorporated into other molecules formed in the plant (Fig. 6-4).

PGAld is the basic molecule used in the formation of organic compounds which become both food and structure for plants. These compounds are the food materials passed on to other organisms as well. PGAld has a relatively short lifetime, as it gets incorporated almost immediately into more complex molecules. The most common conversion of PGAld is into a simple sugar, $C_6H_{12}O_6$. The usual representation of photosynthesis generally denotes this conversion. However, many different compounds are formed from the intermediary PGAld.

The dark reactions, then, involve the actual enzymatic syntheses of the organic chemical food-fuels which the biological community

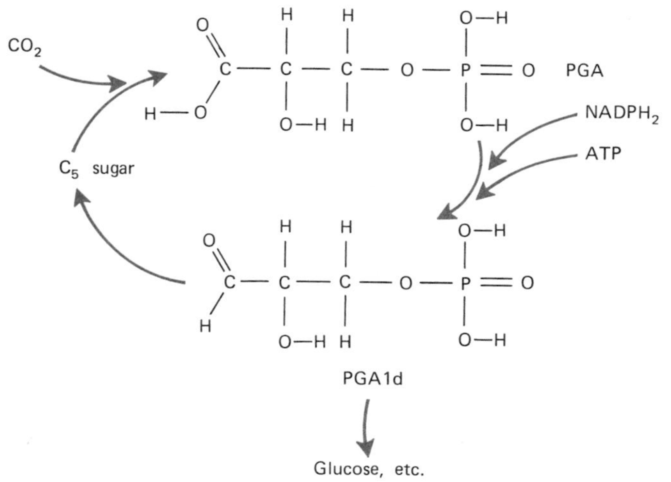

6-4

A highly schematic diagram of CO_2 fixation.

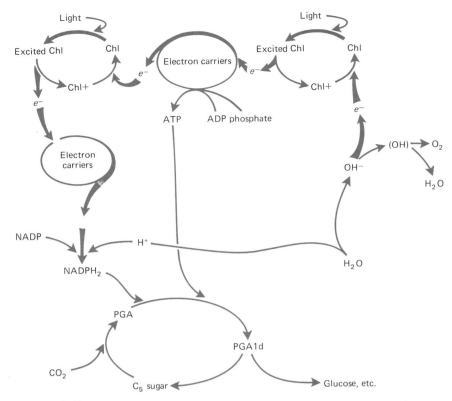

6-5 A schematic diagram of photosynthesis. The heavy line denotes electron flow.

depends upon for its suitable constant energy source. The dark reactions require CO_2 in the atmosphere and the energy-rich molecules $NADPH_2$ and ATP formed in the light reactions. The light reactions require light, chlorophyll, water, and the necessary enzyme and electron transport systems to produce these energy-rich compounds (Fig. 6-5).

Suggested Readings

ARNON, D. I. "The Role of Light in Photosynthesis," in *The Living Cell*, W. H. Freeman, San Francisco, Calif., 1965.

BASSHAM, J. A. "The Path of Carbon in Photosynthesis," in *The Living Cell*, W. H. Freeman, San Francisco, Calif., 1965.

BENNETT, T. P., and EARL FREIDEN, *Modern Topics in Biochemistry*, Macmillan, New York, 1966.

GOLDSBY, R. A. *Cells and Energy*, Macmillan, New York, 1967.

Respiration 7

Chemical reactions occur within a living organism which maintain the structure and function of the organism. These reactions, however, require an input of "biologically usable" energy. The most available form of energy within the cell is the chemical energy in the structure of ATP. ATP is synthesized within the cell during the stepwise decomposition of organic molecules. This decomposition gives off the energy stored in the organic foodstuffs during their synthesis, and this energy then drives the synthesis of ATP from ADP and inorganic phosphate. By this energy transfer, the cell is provided with a ubiquitous energy source which can be used in almost all cell processes. Photosynthesis is the ultimate source of the organic molecules which are decomposed to form ATP. Hence the formation of ATP is the final step in the transformation of light energy into chemical energy of a biologically useful form.

The most efficient cellular process by which ATP is formed during the breakdown of a wide variety of organic compounds requires molecular oxygen, O_2. This process is called respiration, and the overall chemical equation is

$$6\ C_6H_{12}O_6 + 6\ O_2 \longrightarrow 6\ CO_2 + 6\ H_2O + \text{energy}$$

Respiration requires a gas exchange between the cell and its surroundings to allow for the inflow of O_2 and the outflow of the CO_2 produced as a breakdown product. Respiration is strictly the oxygen-requiring, or aerobic, process of ATP production. This biochemical meaning of respiration should not be confused with its everyday meaning of breathing or the processes of ATP production which occur in some organisms which do *not* involve O_2.

The most common substance decomposed aerobically in cells to produce ATP is glucose, $C_6H_{12}O_6$. The breakdown of glucose will be described in detail to indicate the mechanism of ATP production within the cell. The scheme for glucose can then be related to the breakdown of other organic compounds to form ATP. Although the breakdown of glucose is a continuous process, it can be discussed in three "steps." The first step, glycolysis, is anaerobic; it does not require O_2. The next two steps are part of the respiratory sequence and require O_2. They are the Kreb's citric acid cycle and electron transfer. The anaerobic step, glycolysis, takes place in the cytoplasm. The aerobic steps occur in the mitochondria.

Glycolysis

The process of glycolysis is common to the aerobic breakdown of glucose and to the two different types of *an*aerobic breakdown of glucose. One type of anaerobic glucose decomposition, fermentation, occurs in yeast. The other type occurs in muscles. In the overall process of glycolysis the C_6 sugar, glucose, is slowly broken down in various enzymatic steps to two C_3 units of pyruvic acid (Fig. 7-1). The subsequent reactions of this common glycolytic product, pyruvic acid, will be discussed separately under respiration, fermentation, and anaerobic ATP production in muscles.

The first step of glycolysis is the addition of phosphate to glucose. This phosphorylation requires energy, using an ATP. The glucose phosphate changes readily to another C_6 sugar phosphate, fructose phosphate. In another ATP-requiring reaction, fructose phosphate is phosphorylated, adding another phosphate, to form fructose diphosphate. This C_6 unit cleaves to form two C_3 units of phosphoglyceraldehyde (PGAld). The PGAld is oxidized by the removal of two electrons and two H^+ ions to form phosphoglyceric acid (PGA). The hydrogen carrier NAD accepts these electrons and H^+ ions to become its reduced form, $NADH_2$. This reaction releases energy which drives the synthesis of ATP from ADP and phosphate. The PGA formed also possesses a high energy $\sim PO_4$ bond. In several steps this compound changes to pyruvic acid. During this transformation, the $\sim PO_4$ group is transferred to ADP to form ATP, thus trapping the energy stored in the PGA.

In the glycolytic breakdown of a molecule of glucose two ATPs are used up and four ATPs are produced. The other products are two $NADH_2$ and two pyruvic acids. The net gain of ATPs during glycolysis is two ATPs per glucose.

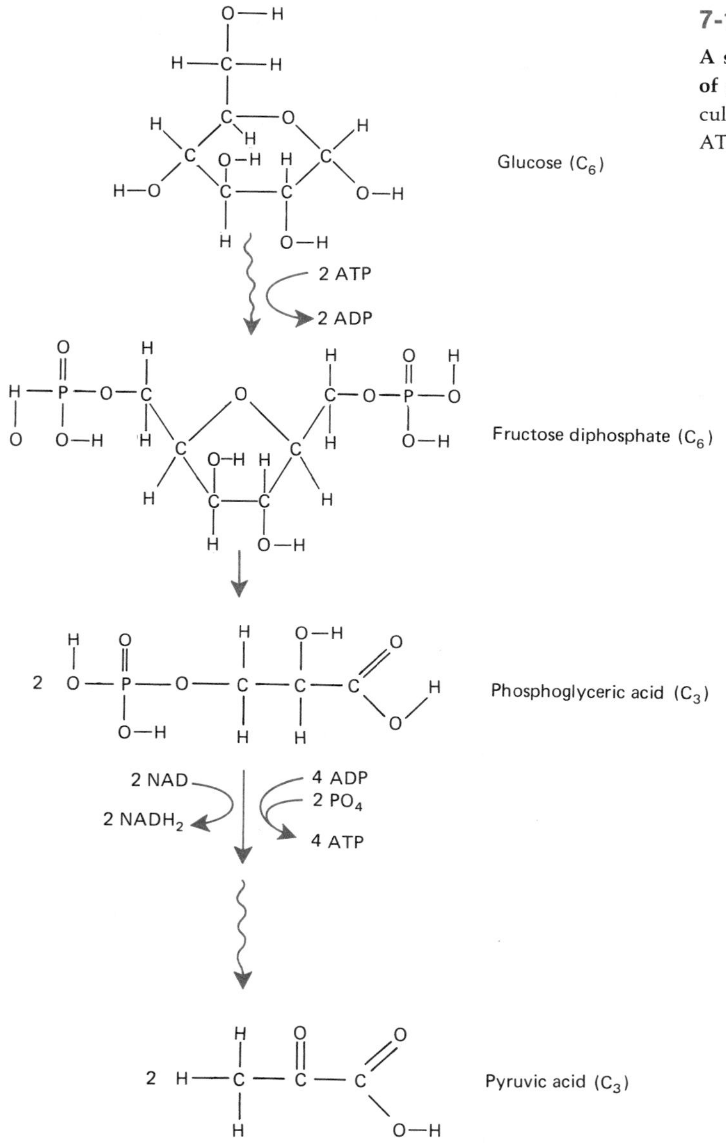

7-1

A simplified version of the basic process of glycolysis. During glycolysis one molecule of glucose is broken down to two ATP, two NADH$_2$, and two pyruvic acids.

Krebs' Citric Acid Cycle

In the presence of O$_2$ the pyruvic acid formed in glycolysis is broken down further in the mitochondria. The pyruvic acid loses CO$_2$ to the atmosphere, loses two e^- and two H$^+$ to NAD, and reacts with the enzyme cofactor coenzyme A, CoA, form the com-

Respiration [101]

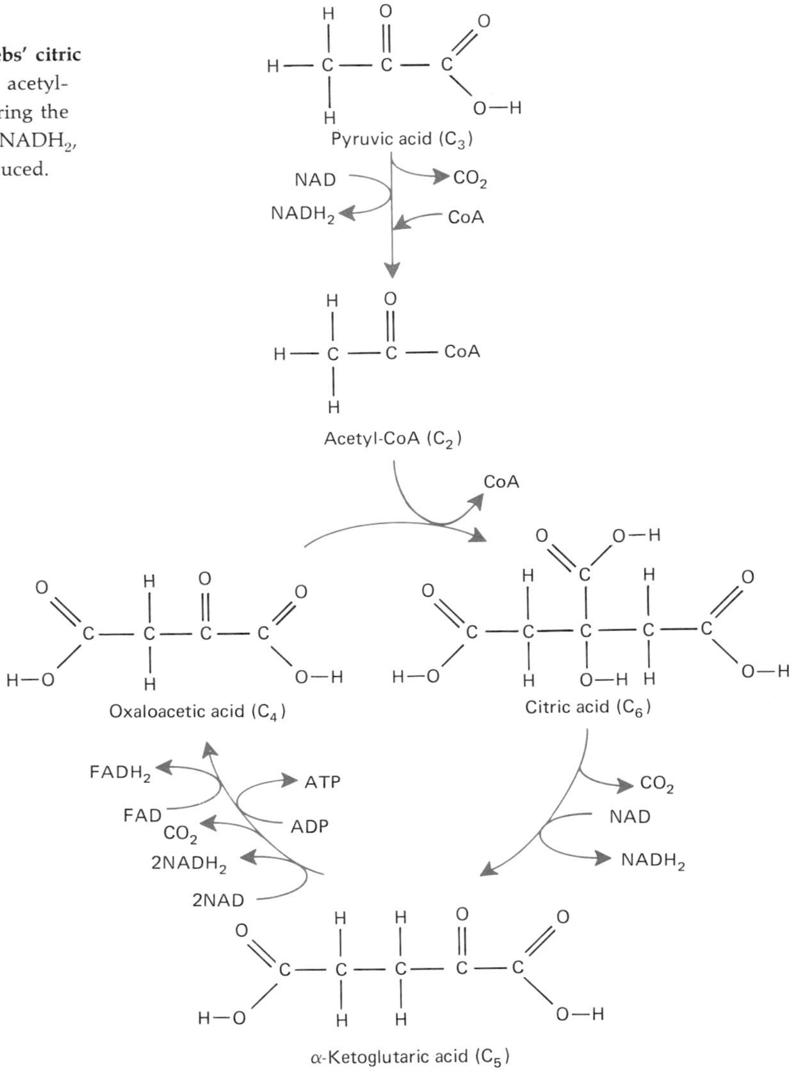

7-2

A simplified version of the Krebs' citric acid cycle. For every C_2 unit of acetyl-CoA which is broken down during the cycle there are two CO_2, three $NADH_2$, one $FADH_2$, and one ATP produced.

plex acetyl-CoA, CH_3CO—CoA. This complex then reacts with a C_4 acid, oxaloacetic acid, which is present in the mitochondria. The reaction forms a C_6 unit, citric acid, and regenerates CoA. Citric acid enzymatically undergoes a series of stepwise changes which ultimately lead to the regeneration of oxaloacetic acid (Fig. 7-2). Because this oxaloacetic acid can then react with another acetyl-CoA formed by pyruvic acid breakdown and start the process all over again, this sequence of reactions is called the Krebs' citric acid cycle. It is named for Sir Hans Krebs who first postulated the scheme in 1937.

The electron transfer cofactors play an important role in the reactions of the citric acid cycle. The citric acid initially formed from acetyl-CoA and oxaloacetic acid changes in steps to another C_6 acid. This acid loses CO_2 and loses two e^- and two H^+ to NAD to form a C_5 acid, ketoglutaric acid, and $NADH_2$. This C_5 acid also loses CO_2 and, in a series of subsequent steps, loses six e^- and six H^+ to two NADs and one FAD to regenerate the C_4 oxaloacetic acid. An ATP (or its equivalent, a guaninetriphosphate) is formed during this sequence of transformations.

In toto, from every pyruvic acid that enters the citric acid cycle there are three CO_2, four $NADH_2$, one $FADH_2$ (the reduced form of FAD), and one ATP produced.

Electron Transfer Systems

Most of the ATP produced during respiration is produced in the electron transfer systems. The two $NADH_2$ produced in glycolysis, the two $NADH_2$ produced in the acetyl-CoA formation, and the six $NADH_2$ and two $FADH_2$ produced in the citric acid cycle all donate the electrons which they captured in these reactions to the enzyme systems in the mitochondria which contain electron carriers. Each electron carrier has a slightly different electron potential. As the electrons from the cofactor $NADH_2$ get transferred from one electron carrier to the next, they slowly give up their energy. This energy is used in the energy-requiring synthesis of ATP from ADP and inorganic phosphate. The electrons are ultimately donated to O_2 which then picks up protons to form H_2O. There are several kinds of electron carriers which participate in this process. It is not known with certainty whether the protons are transferred along with the electrons in this transport or not. The currently accepted scheme, however, is shown in Fig. 7-3.

7-3 **A schematic diagram of the electron transport system.** Each pair of electrons which gets transferred to O_2 triggers the formation of three ATPs.

This sequence illustrates why the breakdown of glucose requires O_2. Oxygen serves as the ultimate electron acceptor for the electrons captured by cofactors during glucose decomposition. Furthermore, in the process of electron transport from $NADH_2$ to O_2, each pair of electrons causes the stepwise formation of 3 ATP units. One ATP unit is formed during the first step of electron transfer from $NADH_2$ to FAD. During the subsequent transfer from $FADH_2$ through the cytochrome system to O_2 two more ATP units are formed. Hence when electrons are donated directly to FAD in the citric acid cycle, only 2 ATP units are subsequently formed during electron transport.

ATP Production during Glycolysis and Respiration

The net products from glycolysis are 2 ATPs and 2 $NADH_2$ per glucose molecule. Since each $NADH_2$ molecule leads to 3 ATPs during electron transport, a total of 8 ATPs results from glycolysis and electron transport.

In the citric acid cycle 4 $NADH_2$, 1 $FADH_2$, and 1 ATP are formed during the breakdown of each pyruvic acid. Each glucose molecule, then, (which gives 2 pyruvic acids) leads to the formation of 8 $NADH_2$, 2 $FADH_2$, and 2 ATPs in addition to the glycolysis products. The number of ATPs formed during the citric acid cycle and electron transport, then, is 24 + 4 + 2 or 30. In total, 38 molecules of ATP are synthesized during the complete breakdown of one molecule of glucose. This represents a capture of about 60% of the energy available from the breakdown of glucose. This is a fairly high efficiency compared to that of other machines.

Fermentation

Fermentation is the process by which yeast breaks down glucose anaerobically. The products of fermentation are CO_2, ethyl alcohol (CH_3CH_2OH), and ATP. In yeast, glucose breaks down, as in glycolysis, to produce two pyruvic acids, two ATP, and two $NADH_2$. The pyruvic acid, in an enzymic reaction, breaks down into CO_2 and a C_2 compound, acetaldehyde (CH_3CHO). Since the entire process occurs without oxygen, the $NADH_2$ does not give its electrons to O_2 through the electron transport system as it does

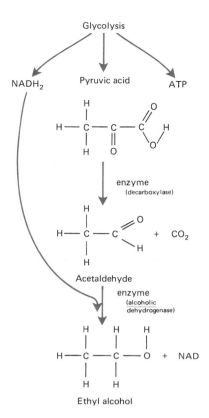

7-4

The chemical processes occurring in fermentation.

in respiration. Instead, the $NADH_2$ donates its two e^- and two H^+ to the acetaldehyde in another enzyme-catalyzed reaction. This regenerates the NAD and forms ethyl alcohol (Fig. 7-4). The NAD, then, can be used again in the glycolytic portion of glucose decomposition.

The total fermentation process only leads to the formation of two ATP per glucose molecule processed. This energy-capture mechanism is much less efficient than respiration.

Anaerobic Production of ATP by Muscles

ATP is also produced anaerobically in human muscles. Again, this process starts with glycolysis, but the pyruvic acid formed suffers a slightly different fate. The glycolytic portion nets two pyruvic acids, two ATP, and two $NADH_2$ per glucose molecule. As in fermentation the $NADH_2$ cannot donate its electrons to O_2. Instead,

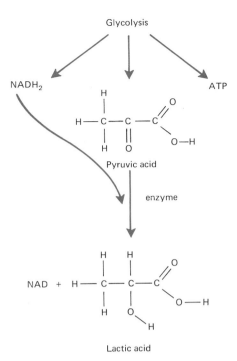

7-5

The chemical processes occurring in muscles without a sufficient O_2 supply.

the $NADH_2$ donates them to the central atom of pyruvic acid to form lactic acid (Fig. 7-5). The formation of lactic acid is what causes soreness in muscles that are overworked. When muscles are worked hard, the muscle cells need to produce extra energy in the form of ATP. Respiration produces the most energy. If, however, the muscle is worked faster than O_2 can be supplied to it from the bloodstream, the muscle will produce ATP anaerobically and lactic acid accumulates. The formation of lactic acid regenerates the NAD for use in the glycolytic sequence.

The anaerobic formation of ATP by muscles is also much less efficient than the aerobic respiratory process. Only two molecules of ATP are produced per molecule of glucose.

Production of ATP from General Food Compounds

The production of ATP by the respiratory breakdown of glucose is the most common source of ATP in higher plants and animals. In fact, the decomposition of most organic compounds, carbohydrates, proteins, or fats, proceeds through the Krebs' cycle. All these compounds, then, must enter this glycolysis–Krebs' cycle–electron transport sequence at some point. The form in which these complex foodstuffs enter the sequence depends upon their structure and the preliminary reactions which transform them into suitable participants of this sequence. These processes are summarized in Fig. 7-6 to show how the different foods are used within the human body to produce ATP.

Carbohydrates, in general, enter the respiratory sequence as glucose units. Intermediate reactions are required to break down most carbohydrates into glucose units. Disaccharides, such as sucrose, are broken down during digestion to simple sugars which are then taken into the body. These simple sugars are readily converted to glucose and can then enter the glycolytic sequence. Glycogen, which was synthesized by the cell from glucose, can be converted back to glucose. Starch, too, can be converted back to

7-6 The pathways by which various foods and products of their breakdown in the body enter the glycolysis-Krebs' cycle-electron transport sequence which produces ATP.

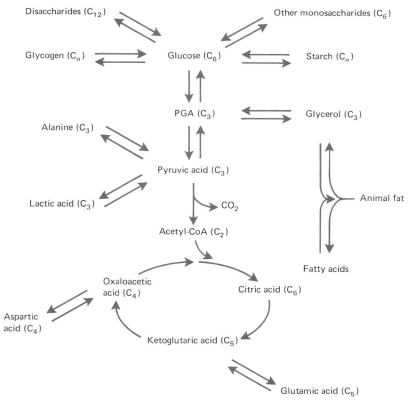

glucose units for respiration. By the conversion of these carbohydrates into glucose, carbohydrates are broken down into CO_2 and H_2O in the manner already described. This is the primary means of ATP production in higher plants and animals.

Lactic acid, formed in muscles during anaerobic break down of glucose from pyruvic acid, gets converted back to pyruvic acid and enters the respiratory sequence there. Glycerol, the C_3 trialcohol formed in the break down of animal fat, is converted to PGAld by the cell. Glycerol, then, enters the respiratory sequence at the PGA stage of glycolysis, is subsequently converted to pyruvic acid, and proceeds through the citric acid cycle from there.

Fatty acids and some amino acids are broken down into acetyl groups which enter the citric acid cycle as the acetyl-CoA complex. Free fatty acids are activated by CoA and then oxidized by FAD and NAD in the presence of water to produce acetyl-CoA units. This series of reactions can be briefly summarized as follows:

$$R-CH_2-CH_2-CH_2-COOH \xrightarrow[ATP]{CoA}$$

$$R-CH_2-CH_2-CH_2-\overset{O}{\underset{\|}{C}}-CoA \xrightarrow[\substack{H_2O \\ CoA}]{\substack{NAD \\ FAD}}$$

$$R-CH_2-\overset{O}{\underset{\|}{C}}-CoA + CH_3-\overset{O}{\underset{\|}{C}}-CoA$$

The remainder of the fatty acid can then recycle for the removal of another acetyl-CoA unit, and the acetyl-CoA units are broken down further in the Krebs' cycle.

Most of the common amino acids from food protein can be broken down in this respiratory sequence to produce ATP. Since all the amino acids differ in structure, they enter the respiration pathway at different points. For example, the C_3 amino acid, alanine, is converted to pyruvic acid by removal of the amino group (deamination) and other small changes. Hence, alanine enters the respiratory sequence at the pyruvic acid stage. The C_5 amino acid, glutamic acid, however, can be converted to ketoglutaric acid by deamination and other changes. It, therefore, enters the respiratory sequence at that level. The C_4 amino acid, aspartic acid, is converted to oxaloacetic acid and is used in respiration at that point. Other amino acids require other conversions, but most of them can ultimately be broken down through respiration to produce ATP.

Suggested Readings

BENNETT, T. P., and EARL FRIEDEN, *Modern Topics in Biochemistry*, Macmillan, New York, 1966.

GOLDSBY, R. A. *Cells and Energy*, Macmillan, New York, 1967.

JELLINCK, P. H. *The Cellular Role of Macromolecules*, Scott, Foresman and Company, Glenview, Ill., 1967.

LEHRINGER, A. L. "How Cells Transfer Energy," in *The Living Cell*, W. H. Freeman, San Francisco, Calif., 1965.

8 The Movement of Substances Into and Out of Cells

All cells are surrounded by a cell membrane composed of proteins, carbohydrates, and modified fats. This membrane must keep within the cell those things needed in cells. The membrane must also allow substances in the cell's environment which a cell requires to function and the waste substances which a cell produces to pass into and out of the cell, respectively. There are three basic mechanisms by which this movement occurs, diffusion, active transport, and endocytosis. Diffusion describes the movement of substances through any medium which results from the random motion of the particles of that substance. The principles of diffusion will be discussed in detail and it will become evident that some substances move differently in biological systems than these principles predict. Cells can use energy to alter the movement of substances through the cell and its membrane.

Principles of Diffusion

Diffusion is the net directional movement of a substance through a medium which results from the random motion of the particles of that substance. The particles of the substance under observation may be molecules or ions dissolved in a medium or, in some cases, colloidal particles suspended in a medium. For simplicity only molecular diffusion will be discussed, although the same principles apply to all other types of diffusion.

A molecule is in constant motion, and its direction of movement is completely random. It can go in any direction at any time. When two molecules collide, their direction of motion changes. A typical path of a molecule (represented in only two dimensions) would look something like this:

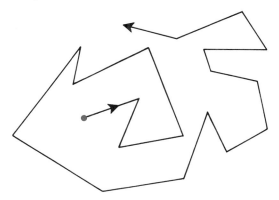

This pattern is probably representative of molecular motion in either a liquid or gaseous medium. Each path length, however, would be longer on the average for a molecule in a gas than for one in a liquid since molecules are farther apart in a gas and collisions are thereby less frequent. Molecules in a solid do not move from certain average positions within the solid. They merely vibrate or "rattle around" in that fixed position.

When molecules are introduced into a small portion of a medium in which they can move, either liquid or gaseous, diffusion occurs. That is, molecules of that substance will move throughout the medium causing a net directional motion away from the point of introduction. This diffusion results directly from the random and constant motion of molecules. If there is a high concentration of molecules at point A in a medium and none at point B, the random motion of these molecules in the medium will cause some to move from A toward B. However, there are none at B to move back to A. The net directional movement will, therefore, be from A to B until the molecules are equally distributed throughout the medium (Fig. 8-1).

Diffusion, then, only occurs when a concentration gradient exists within a medium. That is, the concentration of a substance is different at one point in the medium than it is at another. The net movement, or diffusion, is always from the higher concentration area to the lower.

The rate of diffusion depends in part on the velocity of the molecules involved. The faster the molecules move, the faster will the net movement from higher to lower concentration occur. Diffusion generally occurs much faster in a gaseous medium than in a liquid medium because molecules move faster in gases than they do in liquids. Diffusion effectively does not occur in a solid medium

8-1

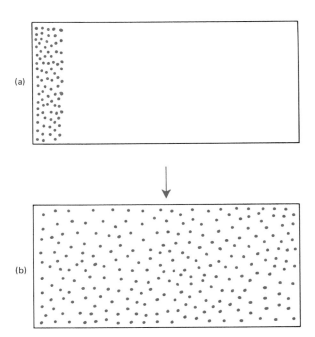

The process of diffusion. (a) Since molecules are in random motion, some of them will move away from the concentrated area. There are, however, none present elsewhere in the medium to move toward the concentrated area. Therefore, the net molecular movement will be from left to right. (The molecules of the medium are not shown; their movement averages out.) (b) Once the molecules are evenly distributed through the medium, all further molecular motion is essentially equal in all directions. Therefore, no net change occurs.

because molecules do not move from a given relative position in a solid. Within a given physical state, the velocity of molecules increases as the temperature rises. Diffusion occurs faster, therefore, in hot water than it does in cold.

The rate of diffusion also depends on the sharpness of the concentration gradient. Although the movement of molecules at any given temperature is always the same, the *net* movement slows down as the concentration gradient decreases. When a large concentration gradient exists, more molecules are moving toward the lower concentration area than are moving away from it because there were not very many there in the first place. As diffusion proceeds, the concentration increases in the area where the concentration was originally low. Now there are more molecules there to move away to balance the number of molecules moving toward the area. The rate of diffusion, therefore, slows down, although the molecular rate of motion does not change.

Diffusion Through a Membrane

The process of diffusion can take place through a membrane if the membrane allows passage of the diffusing substance. A membrane is permeable to a molecule or ion if that particle can pass through the membrane. A differentially permeable or semipermeable membrane is one that some particles can pass through faster than others can. The diffusion of molecules in a system with a permeable membrane occurs through the same mechanism of random movement as if there were no membrane. The membrane may slow down the rate of diffusion, but the mechanism of diffusion is the same.

Suppose a membrane separates a solution of a substance in a solvent from the pure solvent. If that membrane is equally permeable to both solvent and solute, these two substances will diffuse across the membrane according to their concentration gradients. Since the concentration of solvent is highest in the pure solvent, it will diffuse into the solution. Likewise, the concentration of solute is highest in the solution; therefore solute will diffuse into the solvent. These trends will continue until the solute is evenly distributed throughout the entire system (Fig. 8-2). If, however, the

8-2 **Diffusion through membranes and the process of osmosis.** (a) The net movement of solvent molecules is into the solution. The net movement of solute molecules is into the solvent. These trends continue until the solute is evenly distributed throughout the entire system. (b) The solute does not cross the impermeable membrane. The net movement of solvent is into the solution. This movement necessarily builds up pressure, called osmotic pressure. This net movement will continue until the pressure is such that just as many solvent molecules diffuse into the solution as move out.

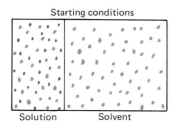
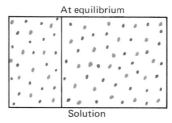

(a) Solution is separated from pure solvent by a membrane which is permeable to both solute and solvent

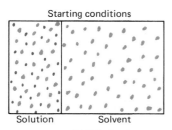
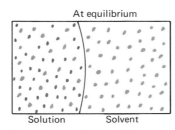

(b) Solution is separated from pure solvent by a membrane which is permeable only to solvent

membrane is permeable to the solvent but not to the solute, the solute cannot diffuse into the solvent. Only the solvent can diffuse into the solution chamber, where its concentration is lowest. This process is called osmosis.

Movement Through a Cell Membrane

There are three mechanisms by which materials move into and out of cells: diffusion, active transport, and endocytosis.

Diffusion Through Cell Membranes

Cell membranes are not equally permeable to all compounds which pass into and out of the cell. Cell membranes are freely permeable to H_2O, CO_2, and O_2, and these substances move into and out of the cell by diffusion. CO_2 can dissolve in water to form H_2CO_3 or HCO_3^- and H^+. Both H_2CO_3 and HCO_3^- also diffuse freely into and out of cells. The concentration gradient for CO_2 and O_2 movement is provided by the respiratory systems in the cells. As O_2 is used up in the respiratory sequence, there is a higher O_2 concentration outside the cell than inside; therefore O_2 diffuses in. Simultaneously, as CO_2 is produced in this sequence, there is a higher concentration of CO_2 inside the cell than outside; therefore CO_2 diffuses out. Since cytoplasm is mostly water with dissolved substances, the concentration gradient for water would be from pure water into the cell. However, the fluids outside the cells of multicellular organisms exist in general concentration balance with the cytoplasm. Were this not so, water would continuously diffuse into the cell, possibly causing its rupture.

Active Transport

Active transport is the process by which materials move into a cell against a concentration gradient. That is, substances are moved across a membrane from an area of lower concentration into one of higher concentration. This process requires energy. Most ions are moved through cell membranes by active transport because many of them are chemically incompatible with the cell membrane. The cell membrane is also relatively impermeable to other necessary materials, such as glucose. Such impermeable substances generally move into the cell by active transport.

The mechanism of active transport is not completely understood. The process requires an energy source, which is presumed to be

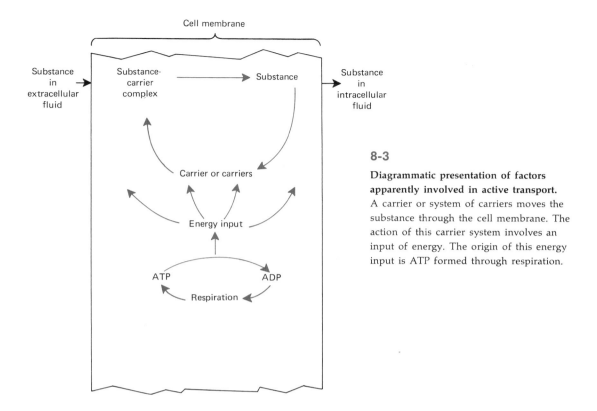

8-3

Diagrammatic presentation of factors apparently involved in active transport. A carrier or system of carriers moves the substance through the cell membrane. The action of this carrier system involves an input of energy. The origin of this energy input is ATP formed through respiration.

ATP. In the most generally accepted theory of active transport, the material to be moved complexes with a carrier substance or system of such carriers within the cell membrane. These carriers move the material through the cell membrane. Energy must be supplied for this carrier system to function (Fig. 8-3).

Active transport is an important process to the cell for many reasons. Because it allows movement of materials against a concentration gradient, albeit with an expenditure of energy, active transport allows an organism to "stockpile" waste products for efficient removal from the organism. A good example of this phenomenon is the formation of urine in the human kidney. The kidney, by active transport, is able to concentrate the important body waste product, urea, to a concentration that is 60 times what it is in the blood which transports it to the kidney. Active transport also affords the movement of substances into the cells which cannot effectively diffuse through the membrane. Many ions which are necessary to cell function, such as Na^+, K^+, Cl^-, Ca^{2+}, Mg^{2+}, and H^+, are generally moved through the cell membrane by active transport. Finally, since active transport requires energy, it is controlled by the energy

The Movement of Substances Into and Out of Cells

sources within the cell. The more passive process, diffusion, is not under such control. Active transport allows the maintenance of a certain balance of some materials inside the cell. By maintaining a delicate balance through active transport, cell function can be maximized.

Endocytosis Some substances move into the cell if the cell membrane invaginates around them and pinches off to form a vacuole inside the cell. The membrane of this vacuole, the original cell membrane of the invagination, then breaks down and the contents of the vacuole are released into the cell. This general process is referred to as endocytosis. If only fluids or solutions are engulfed in this manner, the process is called pinocytosis. The process in which particles, such as food particles, are absorbed in this manner is called phagocytosis.

Pinocytosis is an important means of material transport for many cells as some proteins and hormones apparently enter the cell by this process. These molecules are evidently too large and complex in structure to enter the cell through the membrane either by diffusion or active transport. The mechanism of the uptake of these proteins and hormones apparently involves contact between the cell membrane and the substance being transported. Once this contact is established, the membrane impouches around it, pinches together, and releases the compound into the cytoplasm (Fig. 8-4).

8-4 **Diagramatic presentation of one proposed mechanism of pinocytosis.**
A molecule such as a protein (a) binds to the cell (b) and an inpouching of the cell membrane forms (c). This inpouching is pinched off (d) to form a vacuole (e). The breakdown of the membrane around the vacuole leaves the large molecule inside the cell (f).

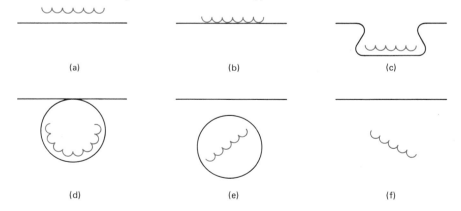

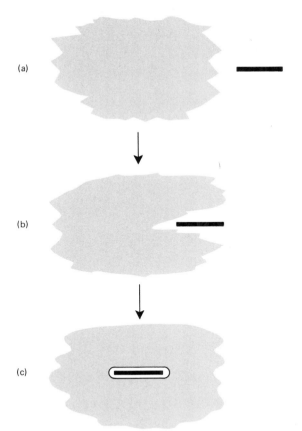

8-5

Diagramatic presentation of the process of phagocytosis. The white blood cell recognizes a bacteria (a), forms an inpouching of the cell membrane around it (b), and closes off the invagination to bring the bacteria inside the cytoplasm of the cell (c).

Phagocytosis is the process by which white blood cells of the human body "digest" the foreign materials in the blood. Once the foreign particle is recognized, the cell membrane surrounds it, pinches off, and forms a vacuole. Usually the foreign substance is then broken down inside the cell (Fig. 8-5).

The cell membrane is impermeable to the cellular macromolecules such as proteins, polysaccharides, nucleic acids, and fats. Once these compounds are formed in the cell, they are confined to its interior by the membrane. They can, of course, be removed by chemical breakdown, but they can be removed intact by reverse endocytosis. Some cells secrete large amounts of protein by reverse phagocytosis. The protein first accumulates in a membrane-bound sac within the Golgi apparatus. This sac moves to the cell membrane and fuses with it. This process releases the contents of the sac outside the cell.

Part III

Genetics

Two fox cubs looking at the world.
(*Courtesy of Colorado Game, Fish, and Parks Division.*)

Chromosomes 9

Genetics is the study of both the transmission and expression of hereditary information. The nucleic acid DNA has been implicated as the carrier of this information in most living organisms. In some viruses, however, RNA is the genetic material. The DNA in the cells of higher plants and animals occurs complexed with proteins. This complex is called nucleoprotein because of its nucleic acid and protein components. The study of genetics in higher organisms necessarily entails a close look at this nucleoprotein complex. The DNA of bacteria and viruses, however, is not found associated with proteins.

The nucleoprotein complexes in the cell are called chromosomes; they are located in the nucleus of the cell. The chromosomes differ chemically from the rest of the nuclear material and can be specifically stained by certain chemicals. Such staining allows their forms and positions within the nucleus to be studied under the microscope. In cells not undergoing division, the chromosomes appear as fine strands in the nucleus. In dividing cells, however, they appear as short, thick strands. Although the term chromosomes is the proper term for both forms, the fine strands found in the resting nucleus are sometimes referred to as chromatin.

Since chromosomes in many organisms are easily seen under the microscope, the behavior of this hereditary material can be observed during all stages of cellular activity. Organisms display a general continuity of form; therefore an accurate transfer of genetic information must occur in cell division. A study of chromosomal behavior uncovers the mechanism of this information transfer. During cell division chromosomes themselves replicate, and the replicated chromosomes are distributed between the two daughter cells in a very specific way. The characteristics of chromosomes, such as how they replicate, how they occur in the nucleus, and how they change, will be discussed in this chapter. The distribution of the replicated chromosomes into the daughter cells will be described in the following chapter.

During this discussion of the *inheritance* of the genetic material, it is useful to remember how this information is *expressed*. The DNA in the chromosomes represents the codons, genes, and operons discussed earlier. The primary mechanism of expression of the information in a DNA molecule is the control of protein synthesis.

There was a time when the molecular composition and molecular mode of replication and expression of hereditary material was not known. Geneticists merely studied the traits of organisms and their offspring to determine how traits were inherited. By correlating studies of inherited traits and findings about chromosomal behavior, modern geneticists were able to pin-point areas along a chromosome which were apparently responsible for determining particular traits. These loci which governed traits were called genes. A chromosome, then, became a string of genes. This conception of a chromosome is still useful even though our knowledge of its composition and mechanism of expression has grown immensely.

Chromosomal Replication

Chromosomes replicate during the process of cell division. This chromosomal duplication must include the template replication of the DNA content of these nucleoprotein complexes. In this process, discussed in Chapter 4, each DNA molecule acts as a template to form two DNA molecules which are identical to the original template molecule. In this way the hereditary material of each chromosome is accurately duplicated. This duplication apparently takes place when the cell is not visibly undergoing cell division. Presumably proteins for the nucleoprotein complex are synthesized during this time also, because at cell division the nucleoprotein complexes condense and appear as duplicate strands. Under the light microscope these duplicate strands, called chromatids, appear to be held together by a body called the centromere. Recent evidence obtained with an electron microscope, however, indicates that *each* chromatid possesses a centromere and that these centromeres are attached to one another to hold the chromatids together. During cell division these centromeres separate, and the chromatid strands become daughter chromosomes (Fig. 9-1).

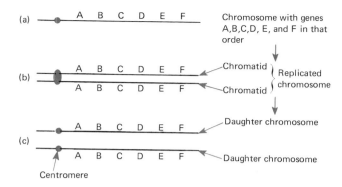

9-1

A diagramatic presentation of the process of chromosome replication. The single chromosome (a) replicates to give a chromosome composed of two chromatids (b). When the centromere divides, daughter chromosomes are produced (c).

Life as it is known today depends on the ability of the chromosome to replicate. This process is essential to growth, reproduction, and the continuity of life forms.

Homologous Chromosomes

The number of chromosomes found in the cells of an organism depends on the organism. In most "higher" plants and animals, the number of chromosomes is an even number. These organisms result from the fusion of two sex cells (egg and sperm), each of which contains a set of N chromosomes. For each chromosome in the set from the father, the paternal chromosomes, there is a corresponding chromosome in the maternal chromosome set. The corresponding chromosome carries genes governing the same traits in the same sequence. Two corresponding chromosomes, one from each parent, compose a homologous pair. For every gene on the paternal chromosome of a homologous pair there corresponds a gene governing the same trait at the same position on the maternal chromosome. These corresponding genes on homologous chromosomes are called alleles. Cells of higher organisms which contain two complete sets of genes are called diploid. Cells with only a single set of genes, such as the sex cells, are haploid.

Chromosomal Mutations

Sometimes certain chemical or physical factors within the cell cause chromosomes to break. The fragments may later recombine. In this process chromosomal segments may get lost, get added onto a different chromosome, or get rearranged within the original chromosome. If a chromosome is permanently changed in this manner and the changed chromosome is replicated, this process is called chromosomal mutation.

A segment of a chromosome may be lost. Radiation by x-rays or atomic radiation has enough energy to cause chromosomes or chromatids to break. A piece may break off the end, leaving a chromosome which is deficient in its terminal genes (Fig. 9-2a). If a chromosome breaks into more than two pieces, a midsegment may be lost when the end fragments recombine without it (Fig. 9-2b). This mutation requires a double break in the chromosome. If a large enough segment is lost from any part of the chromosome, the resulting gene deficiency may be fatal to the organism.

Some chromosomal mutations can lead to a duplication of genes on a chromosome. This happens if chromosomal breakage occurs after chromatid formation. An example of this type of mutation is illustrated in Fig. 9-3. A piece of one chromatid can either get inserted into or added onto the other chromatid of the replicating chromosome. When these two chromatids separate, two altered

9-2

Two examples of chromosomal mutations by deficiencies. The deficiency, in this case caused by breakage and loss of the fragment during cell division can be either terminal (a) or nonterminal (b).

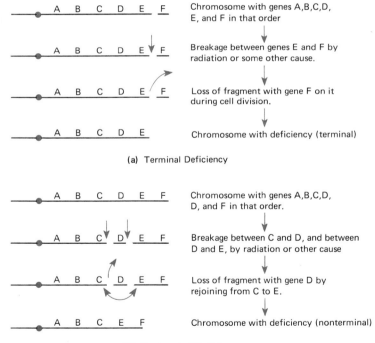

[124] Genetics

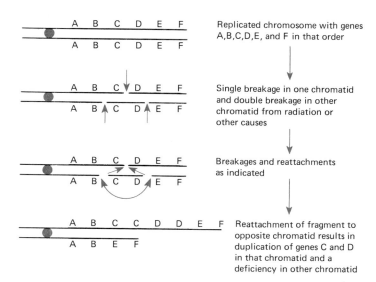

9-3

An example of a mechanism for formation of gene duplications on chromosomes. The double breakage of one chromatid forms a fragment which can be reattached to a single breakage in the other chromatid and result in duplication of genes in that chromatid and in the daughter chromosome it forms upon division of the centromere.

chromosomes form. One chromosome has a gene deficiency, and the other has a gene duplication. Chromosomes with duplicated genes do not necessarily cause a fatal physiological disturbance in a daughter cell.

A terminal fragment lost from one chromosome can recombine with a nonhomologous chromosome. This process is translocation (Fig. 9-4). An inversion can occur if a midsegment formed in a double break recombines with the terminal fragments after turning around (Fig. 9-5). The resulting chromosome contains essentially the same genes, but they appear in a different order than they did on the original chromosome.

The hereditary implications of these chromosomal mutations depends upon the genes involved. A gene deficiency caused by chromosomal deletion can be serious to the resulting organism. Since the genes are missing, the proteins they code are absent. Chromosomal inversion changes the gene sequence of the chromosome. Chromosomal duplication and translocation can result in additional genes on some chromosomes and fewer genes on other chromosomes. These gene alterations can have various effects on the resulting organisms. An extreme example of the result of such a chromosomal mutation is translocational mongolism. This occurs when a large part of the 21st chromosome of a human is translocated to another chromosome. The result of the extra genes in addition to those on the normal homologous chromosome is severe mental retardation and other physiological disturbances.

9-4

Chromosomal mutation by translocation. (a) Two different chromosomes. (b) Breakage occurs in one chromosome. (c) The fragment reattaches to a different chromosome. This process results in two abnormal chromosomes; one has a gene deficiency and the other a gene superfluity.

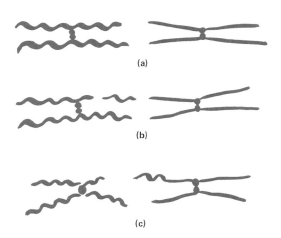

9-5

Chromosomal mutation by inversion. A double break occurs and the midsegment inverts before it recombines with the end fragments.

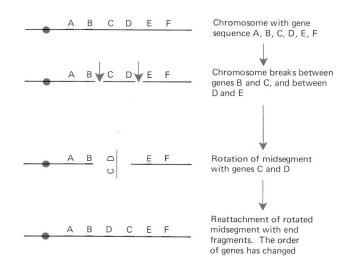

Point Mutations

A mutation which can be traced to a change in a single nucleotide pair in a DNA molecule of a chromosome is called a point mutation. If a nucleotide pair changes and this altered DNA molecule replicates, the hereditary information changes. Since the nucleotide sequence codes protein synthesis, any change in this sequence can interfere with the normal protein synthesis and, by so doing, cell function.

Point mutations have been observed to occur spontaneously. These mutations do not occur often, so it is difficult to determine

[126] Genetics

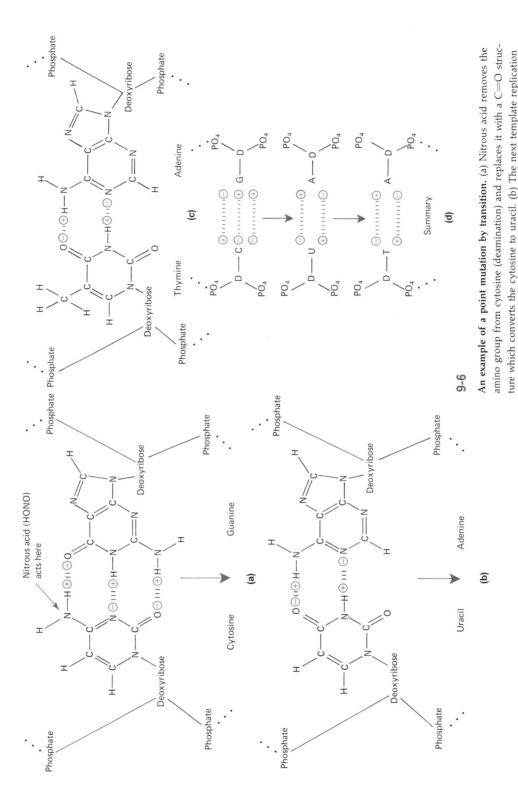

9-6

An example of a point mutation by transition. (a) Nitrous acid removes the amino group from cytosine (deamination) and replaces it with a C=O structure which converts the cytosine to uracil. (b) The next template replication results in the pairing of uracil with adenine. (c) Template replication of DNA replaces uracil with thymine since uracil is not usually included in DNA. The pair is now TA. (d) These steps result in the replacement of the base pair at this location, CG to UA to TA.

[127]

their causes. Most of the point mutations studied have been induced with chemicals or radiation. Spontaneous mutations are no doubt due to similar abnormal, but unidentified, factors within the cell or its environment. In spite of their infrequent occurrence, spontaneous mutations are the ultimate source of all genetic variability. They played an important role in the evolution of the various forms of life.

The mechanism by which high-energy radiation, such as x-rays and atomic radiation, causes point mutation is not well understood. Such radiation can cause components of DNA to cleave and ionize. Presumably, changes in the DNA structure occur as the DNA molecule reforms. Replication of this changed DNA changes the genetic code.

Some chemicals can react with the various components of DNA to cause a point mutation. These chemical agents are mutagens. The study of such chemically induced mutations has widened considerably since the mid-1940's. Environmental and cancer research includes studies of this kind. For example, pesticides are tested for mutagenic action in order to exclude from use any pesticides which apparently cause cancerous type growths.

There are several mechanisms by which chemical mutagens alter DNA. Some chemicals cause a transitional mutation; that is, they can effectively change one nucleotide pair to another. For instance, nitrous acid, HONO, reacts with cytosine to produce uracil (Fig. 9-6). During DNA replication, uracil cannot hydrogen bond to

9-7 An example of a frame-shift mutation. A series of codons (a) can be changed by the insertion of an additional nucleotide into the nucleic acid sequence (b). The shift in the frame of reference results in different codons beyond the insertion, and a change in the genetic information (c).

guanine the way cytosine does. Instead uracil pairs with adenine (Fig. 9-6b). In the next replication the adenine in this new strand pairs with thymine since uracil is not normally incorporated into DNA (Fig. 9-6c). *In toto,* the original CG base pair is changed to a TA pair by an intervening UA pairing.

Some chemicals can cause a nucleotide pair to be deleted from or inserted into the sequence of nucleotides in a DNA molecule. The chemical acridine does this, but its mechanism of action is beyond the scope of this book. Such insertions or deletions of nucleotides change the frame of reference for the codons of the DNA by one base. For instance, a codon series TAGTAGTAG-TAGTAGTAGTAG reads TAG-TAG-TAG-TAG-TAG-TAG-TAG (Fig. 9-7a). Insertion of a single base, C, between two codons shifts the frame of reference so the code becomes TAG-TAG-CTA-GTA-GTA-GTA-GTA (Fig. 9-7c). A nucleotide deletion causes the same type of change. If the second G in the TAGTAGTAGTAGTAGTAGTAG series were deleted, the sequence of codons would read TAG-TAT-AGT-AGT-AGT-AGT-AGT. The triplet combinations beyond the point mutation again change the codon sequence and the genetic meaning.

Suggested Readings

CRICK, F. H. C. "The Structure of Hereditary Material," *Scientific American,* 191, vol. 4, 54–61 (1954).

DEBUSK, A. GIB. *Molecular Genetics,* Macmillan, New York, 1968.

10 Cell Division and the Life Cycle

Cell division is an important process of life. It is essential to the growth and reproduction of living things. For convenience, the process of cell division is arbitrarily divided into three stages. In the first stage the cell prepares for division. During this stage the chromosomes replicate. In the second stage each replicated chromosome separates and moves away from its twin so that two identical sets of chromosomes are formed within the cytoplasm. New nuclear membranes then form around each set. Finally, the cytoplasm divides to form two new cells.

Since the chromosomes contain the genetic information, their equal distribution into the daughter cells is a critical step in cell division. This distribution can be observed under the microscope because the chromosomes of dividing cells can be specifically stained with certain chemicals. Two types of cell division have been observed, mitosis and meiosis. In mitosis the replicated chromosomes of the parent cell are distributed among two daughter cells. In meiosis, however, the replicated chromosomes of the parent cell get distributed into four new cells.

Mitotic Cell Division

In mitotic cell division a single cell divides to form two identical cells in such a way that the set of chromosomes in each new cell is the same as that of the original cell. Mitosis occurs in all organisms except bacteria, blue-green algae, and viruses. Mitosis in plant cells is quite similar to that in animal cells. The small differences will be pointed out where they occur. The illustrations of plant mitosis in this chapter are from cells of an onion root tip. Those of animal mitosis are from whitefish embryo cells. In both plants and animals, mitosis is preceded by an interphase during which the cell prepares for division. The process of mitosis is conveniently divided into four phases for study: prophase, metaphase, anaphase, and telophase. These phases are not distinct stages of mitosis; they are merely steps in a continual process. Either during telophase or directly after it, the cytoplasm divides in a process called cytokinesis.

Interphase

A cell in interphase looks essentially the same as cells not preparing for cell division. Such structures as the cell membrane, nuclear membrane, nucleolus, and so forth, are all intact and appear normal (Fig. 10-1). During this time, however, each chromosome within the naturally appearing cell is being replicated by both DNA and protein synthesis. Interphase can last for a short or a long period of time, for example, from hours to months.

Mitosis

Prophase In prophase a dividing cell becomes visibly different from a nondividing cell. The chromosomes shorten by coiling and thicken into strands which are visible when specially stained. These structures are actually made up of two strands, chromatids, which are attached at their centromeres. In addition, the nucleolus and nuclear envelope disappear.

During prophase in plant cells two systems of thin fibers, called the spindles, form at the two positions where the daughter nuclei will form. These positions are called the poles. The chromosomes, which are scattered in the cytoplasm at the position where the parent nucleus had been visible, become attached to the spindle fibers at their centromeres (Fig. 10-2a).

In animal cell prophase, the nuclear envelope disappears and the chromosomes become visible through coiling, just as they did in plant cell prophase. In addition, however, the centrosome divides to form two centrioles. Fibrous structures appear which are made up of fibers radiating from the centrioles. These fibrous structures are called asters and they move to the poles of the cell. Some of the aster fibers extending from the poles become attached to the chromosomes as spindle fibers (Fig. 10-2b).

Metaphase Metaphase is the stage in which the scattered chromosomes observed in prophase line up across the middle of the

10-1 **Interphase of plant cell mitosis.** Photomicrograph (top), diagram of picture (middle), and instructional diagram (bottom). Note the difference between the actual appearance of the cell and the general form of the instructional diagram. All following illustrations of cell division will be in this style. (*Photomicrograph courtesy of Carolina Biological Supply Company.*)

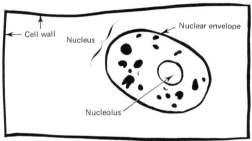

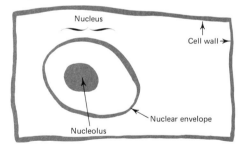

10-2a **Prophase of plant cell mitosis.** Note the formation of chromosomes within nuclear envelope. (*Photomicrograph courtesy of Carolina Biological Supply Company.*)

10-2b **Prophase of animal cell mitosis.** Note the formation of asters and chromosomes. (*Photomicrograph courtesy of Carolina Biological Supply Company.*)

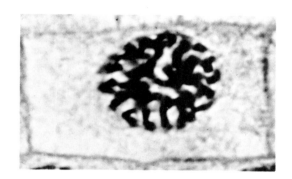
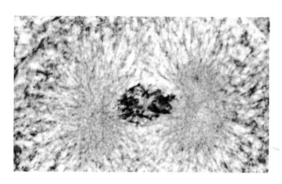

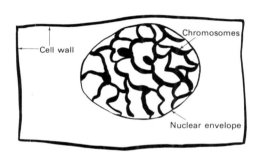
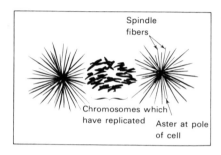

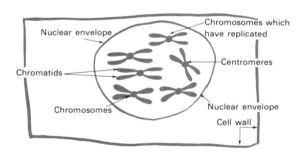
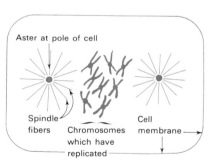

[133]

10-3a **Metaphase of plant cell mitosis.** Note that the chromosomes "line up" across the middle of the cell at a location called the equatorial plane. Note also the formation of the spindle assembly. (*Photomicrograph courtesy of Carolina Biological Supply Company.*)

10-3b **Metaphase of animal cell mitosis.** Note the formation of a line of chromosomes across the equatorial plane of the cell. Note also the clear development of asters and spindle fibers and the attachment of spindle fibers to the chromosomes. (*Photomicrograph courtesy of Carolina Biological Supply Company.*)

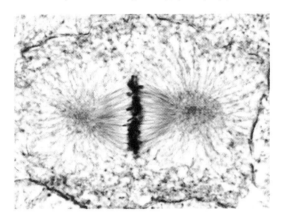
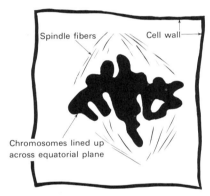
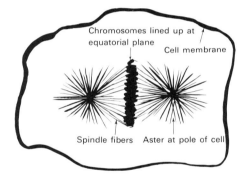
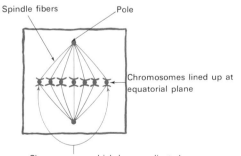
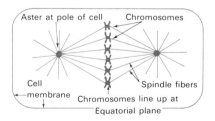

[134]

cell between the two poles. This single line of both homologous and nonhomologous chromosomes across the equatorial plane of the cell is a very distinctive stage of mitosis (Figs. 10-3a, b).

Anaphase During anaphase, the two chromatids which are attached at their centromeres to form a double-stranded chromosome separate and form a pair of identical chromosomes, called daughter chromosomes. One chromosome of each pair moves to each pole. The chromosomes of a cell in anaphase are at various positions between the equatorial plane and the poles. Anaphase ends as the chromosomes reach the opposite poles. At this point, however, fibers still connect the separated daughter chromosomes (Figs. 10-4a, b).

Telophase During telophase a nuclear envelope forms around each group of chromosomes at the poles, and nucleoli reform. Within each new nucleus the chromosomes uncoil, lose their shortened, thickened shape, and revert to the fine strands, chromatin, typical of the interphase cell. In short, the cell at the end of telophase looks much like the interphase cell except that it has two nuclei (Figs. 10-5a, b).

Cytokinesis

The division of the cytoplasm of the plant cell usually occurs during telophase or directly thereafter. A thin wall, the division plate, forms between the chromosomes across the fibers which still connect the daughter chromosomes (Fig. 10-5a). This wall then expands laterally until it extends across the cell, making a thin partition between the two nuclei. The cell wall, typical of plant cells, forms against this partition as the last step in plant cell division. By this process two daughter cells are formed, identical in genetic composition to one another as well as to the original cell.

In animal cells the cytoplasm is constricted across the equatorial plane. By this constriction, the two daughter cells are "pinched off" from one another (Fig. 10-6). In this process two identical cells are formed which are identical to the original cell from which they arose.

Meiotic Cell Division

Meiotic cell division is the type of cell division that produces the gametes, the sex cells, of higher plants and animals. A cell undergoing meiosis divides twice to produce four cells, each with only

10-4a **Anaphase of plant cell mitosis.** This is the phase of movement of the chromosomes to the poles of the cell, and the position of the new cells. (*Photomicrograph courtesy of Carolina Biological Supply Company.*)

10-4b **Anaphase of animal cell mitosis.** The movement of the chromosomes toward the poles of the cell takes place. Note that the chromatids of the chromosomes that have replicated are now separate and called daughter chromosomes. (*Photomicrograph courtesy of Carolina Biological Supply Company.*)

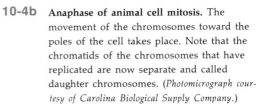

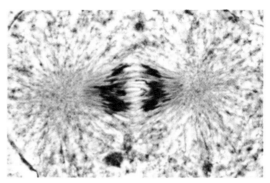

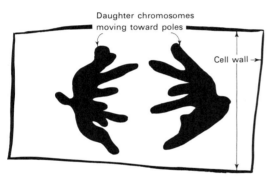

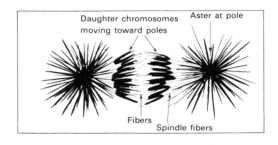

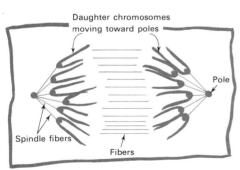

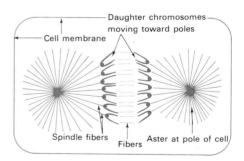

10-5a Telophase of plant cell mitosis. Note the completion of chromosome movement and the reformation of nuclei. Cytokinesis also commonly takes place about this time in the plant cell, but is a separate process from mitosis as such. (*Photomicrograph courtesy of Carolina Biological Supply Company.*)

10-5b Telophase of animal cell mitosis. The movement of the chromosomes is finished and the daughter nuclei re-form. Cytokinesis also takes place at about this time by the formation of a constriction of the cell membrane at the equatorial plane, but this is a separate process from mitosis as such. (*Photomicrograph courtesy of Carolina Biological Supply Company.*)

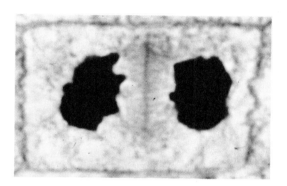

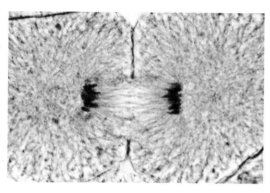

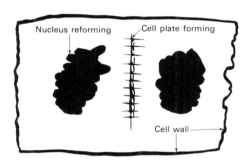

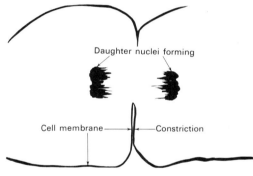

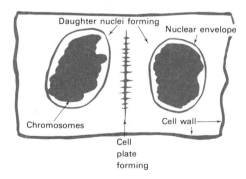

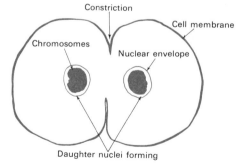

[137]

10-6 The formation of daughter cells in animal cell mitosis. The division of the cytoplasm of the original cell is completed by the constriction, and the nuclei re-form in the new cells. (*Photomicrograph courtesy of Carolina Supply Company.*)

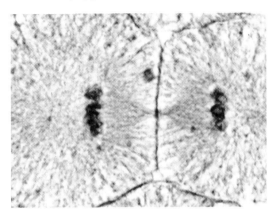

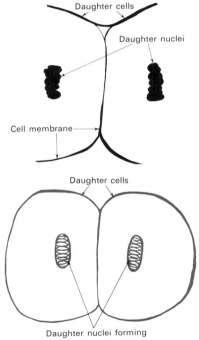

half as many chromosomes as the original cell. Meiosis distributes the two complete sets of chromosomes (2N) of the original cell so that each new cell gets one complete set of N chromosomes. Again, meiotic cell division consists of the three basic cell division stages: interphase or chromosomal replication, chromosomal distribution, and cytokinesis.

Interphase

The preparation of the cell for meiosis is the same as for mitosis. While the chromosomal material replicates, the cell is not visibly different from a nondividing cell (Fig. 10-7).

Meiosis

Many of the phases of meiosis are similar to those of mitosis, so the same terms are used to designate these phases. Since there are two divisions of nuclear material in meiosis, however, Roman numbers I and II are used to designate the meiotic division under study. Although there are two divisions, chromosomal replication only occurs once, prior to the first division. In the first division, the two chromatids of each replicated chromosome do not separate. Instead, the 2N double-stranded chromosomes are merely distributed between the two new nuclei in such a way that one complete set of N chromosomes goes to each new nucleus. The second division more closely parallels a mitotic cell division. At each new nuclear location, the chromatids separate. This process forms four new nuclei, each with N daughter chromosomes.

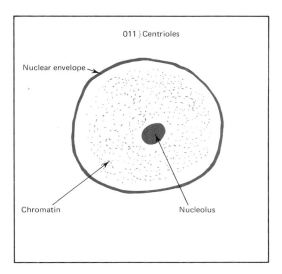

10-7

Cell in interphase that is about to enter meiosis. The chromosomes have doubled to form two chromatids.

10-8

A schematic diagram of the cell in prophase I of meiosis. The chromosomes are now visible, but scattered in the nucleus. Homologous chromosomes pair. (Three different homologous pairs are pictured.) The centrioles replicate and form two asters which migrate to the poles of the cell. The nuclear envelope breaks down by the end of prophase I. (Paternal chromosomes are represented in outline only; maternal chromosomes are shown in black.)

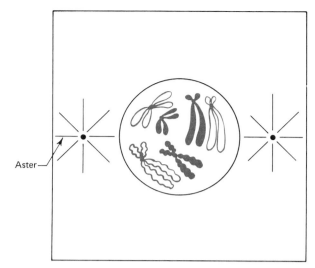

10-9

A schematic diagram of the cell in metaphase I of meiosis. The three different homologous pairs are now lined up at the equatorial plane. One chromosome of each homologous pair is on each side of the equatorial plane.

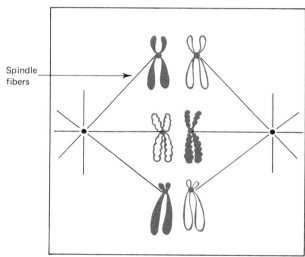

First Meiotic Division PROPHASE I: This stage, similar to mitotic prophase, is characterized by the disappearance of the nuclear envelope and the nucleolus and the appearance of the short, thick chromosomes. As the chromosomes coil, homologous chromosomes pair up. During this pairing homologous chromosomes often come into contact, and chromosomal cross over can occur. At the end of Prophase I the chromosomes are attached to the spindle fibers. These fibers come from the two poles in plant cells and from the developing asters at the poles of animal cells (Fig. 10-8).

METAPHASE I: At this point mitosis and meiosis start to differ. In the meiotic cell the homologous chromosomes pair up. These *pairs* then line up on either side of the equatorial plane to form two rows of chromosomes across the middle of the cell (Fig. 10-9). Because the homologous chromosomes were paired when they lined up, each row contains a complete set of the N different chromosomes. In Figs. 10-8, 10-9, and 10-10, the maternal chromosomes are depicted in black and the paternal chromosomes in outline only. The homologous pairs are randomly aligned so that each row of N chromosomes generally contains both paternal and maternal chromosomes.

ANAPHASE I: As in mitotic anaphase, during anaphase I the chromosomes move toward the poles of the cell. However, in anaphase I the chromatids do not separate into daughter chromosomes. Instead, one set of N chromosomes moves to one pole, and the other set moves to the opposite pole (Fig. 10-10). Only one chromosome from each homologous pair migrates to each pole. Since the pairs were randomly aligned in metaphase I, the set of N chromosomes

10-10 **A schematic diagram of the cell in anaphase I of meiosis.** The maternal and paternal homologous chromosomes separate from one another and move to opposite poles of the cell. The chromatids of the chromosomes do not separate during the first meiotic division. In this division the homologous pairs of chromosomes separate rather than the chromatids of the chromosomes.

10-11 **The cell in telophase I.** Some, but not all species form nuclear envelopes in telophase I. The cytoplasm is not divided during the first meiotic division.

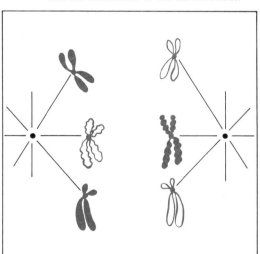

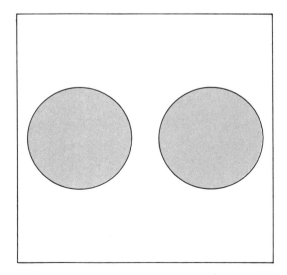

Cell Division and the Life Cycle **[141]**

at each pole is a random combination of maternal and paternal chromosomes.

TELOPHASE I: This stage resembles mitotic telophase, but no cytokinesis accompanies the phase. The chromosomes uncoil and two new nuclei form. In some species the spindle disappears and the nuclear envelope appears, but the cytoplasm is not divided (Fig. 10-11).

Second Meiotic Division In the second division of meiosis, each of the nuclei formed in the first meiotic division divides. No DNA synthesis or chromosomal replication takes place between the first and second meiotic divisions, however. For each nucleus this division is similar to mitotic division, but two nuclei are dividing in the same cell. The resulting cell has four nuclei before cytokinesis.

PROPHASE II: During prophase II the chromosomes become visible through coiling, the nuclear envelope disappears, and spindles form on each side of the daughter nuclei and attach to the chromosomes (Fig. 10-12).

METAPHASE II: As in mitotic metaphase, the chromosomes of each nucleus line up across the equatorial plane between the two poles during metaphase II (Fig. 10-13).

ANAPHASE II: At this point the chromatids separate and the resulting daughter chromosomes move to opposite poles (Fig. 10-14). Assuming no crossover occurred in metaphase I, the chromosomes at the two poles of each nucleus are identical. Of the four new chromosomal groups there are two different sets of two identical chromosome groups. If crossover occurred in metaphase I, the two chromatids of each undivided chromosome of a homologous pair would differ. The resulting daughter chromosomes would not be identical and none of the four new chromosomal groups would be completely identical.

TELOPHASE II: The nuclear envelopes form around the four chromosomal groups and the chromosomes uncoil, giving rise to the diffuse appearance of chromatin in telophase II (Fig. 10-15). In each new nucleus there are only N chromosomes and, if no crossover occurred, there are two sets of two nuclei each with identical genetic material. If crossover occurred, none of the nuclei would carry completely identical genetic information.

Cytokinesis After telophase II the cytoplasm of the original cell divides into the four parts surrounding each of the four nuclei.

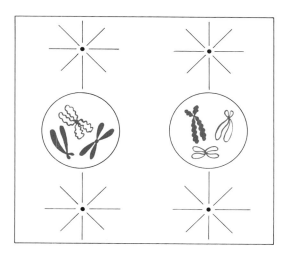

10-12

A schematic diagram of the cell in prophase II of meiosis. The chromosomes again become visible and the spindles reform.

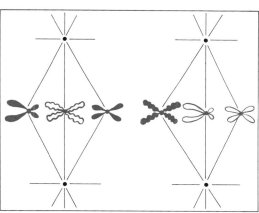

10-13

The cell in metaphase II of meiosis. The chromosomes now "line up" in a single row across the equatorial plane of the cell.

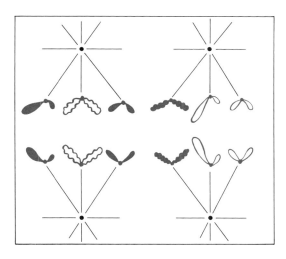

10-14

The cell in anaphase II of meiosis. In this division the chromatids separate to form daughter chromosomes. The daughter chromosomes move to opposite poles.

10-15

The cell in telophase II. The four nuclei form, each with only one set of N chromosomes. The cytoplasm gets divided at the end of telophase to form four cells.

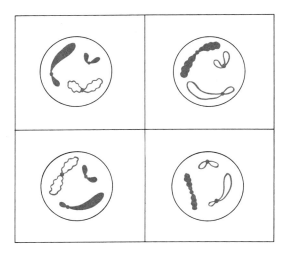

10-16

The basic reproductive cycle involving sexual reproduction typical of most animals.

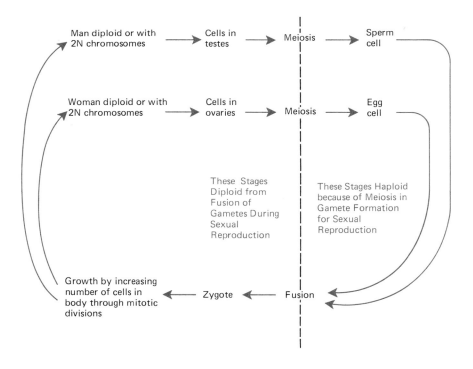

The Role Of Cell Division in Life Cycles

The growth and reproduction of plants and animals are the outcome of cell divisions with mitosis and meiosis. The life cycles typical of "higher" organisms such as vertebrates and flowering plants show the importance of these processes to life.

The life cycle of a human is typical of that of many animals, vertebrates (animals with a backbone), insects, spiders, mites, and so forth (Fig. 10-16). The life cycle is sexual; that is, the fusion of two different haploid sex cells is necessary to reproduce the species. In animals these two different kinds of sex cells are produced by two different types of individuals. These two types of individuals, both diploid, each have certain organs where meiosis occurs to form the haploid sex cells or gametes. The female produces haploid egg cells by meiosis in the ovaries; the male produces haploid sperm cells by meiosis in the testes. Fusion of a sperm and an egg cell gives rise to a fertilized egg cell or zygote. The zygote is diploid, since it results from the fusion of two haploid gametes. The zygote grows from this single cell by mitotic divisions into a sexually mature human. Since all the cells of a human result from mitotic divisions of a single cell, with a few exceptions, the genetic information is identical in every cell, and all cells are diploid. Some genetic heterogeneity can appear within an organism if mitotic crossover occurs. Also, during the process of cell differentiation, some exceptions to this trend of genetic homogeneity arise. For example, haploid cells are produced in the ovary or testes once the new organism achieves sexual maturity. When this occurs the organism is ready for sexual reproduction and another life cycle can begin.

The basic life cycle of flowering plants is different from that of higher animals, but it is also sexual because reproduction requires the fusion of two different gametes. In plants, however, the two different gametes are produced by two different microscopic haploid gametophytes which develop within the plant (Fig. 10-17). The plant, or sporophyte, produces two types of spore mother cells, the megaspore and the microspore mother cells. These cells undergo meiosis to produce megaspores and microspores, respectively. The spores, haploid cells, undergo mitotic division to develop into the mega- and microgametophytes. The megagametophytes produce the female gametes or eggs; the microgametophytes produce the male gametes or sperm. The haploid megaspore develops into the

10-17

The basic reproductive cycle involving sexual reproduction (simplified) typical of the flowering plants, such as trees (except for conifers), flowers, grasses, and bushes.

Sporophytes or "spore producing plant." The tree, flower, grass, or bush that is visible. Diploid or with 2N chromosomes.

→ Megaspore mother cell → Meiosis → Megaspore → Growth by increasing numbers of cells through mitotic cell divisions → Megagametophyte or "female gamete producing plant." → Egg → Fusion → Zygote → Growth by increasing numbers of cells through mitotic cell divisions → (back to Sporophyte)

→ Microspore mother cell → Meiosis → Microspore → Growth by increasing numbers of cells through mitotic cell divisions → Microgametophyte or "male gamete producing plant." → Sperm → Fusion

These Stages Diploid from Fusion of Gametes During Sexual Reproduction

These Stages Haploid because of Meiosis in Gamete Formation for Sexual Reproduction

[146]

haploid megagametophyte at the base of the flower of the plant. The pollen grain, or haploid microgametophyte, develops a pollen tube and two sperm cells after it falls on the sugary secretions of the flower's stigma. The sperm in the pollen tube reaches the egg inside the flower and fertilizes it there to form the diploid zygote. The zygote divides by mitotic cell divisions and develops into the fertile seed. Once it finds a suitable environment, the seed develops into the flowering plant by further mitotic divisions. This development of the plant, or sporophyte, from the zygote is completely mitotic and all cells in the organism are diploid until the two spore mother cells undergo meiosis.

The life cycles of both plants and animals indicate that mitosis is important to growth. All higher organisms grow by the exact genetic duplication of an original zygote. Cell division with mitosis results in the formation of identical cells, each with the same number of chromosomes. Cell division with meiosis does not result in a multiplication of the same cell. During meiosis the chromosomal number per cell is reduced by half. For instance, a series of meiotic divisions starting with a cell with 64 chromosomes would lead to the following cellular arrays:

1 cell of 64 chrom. $\xrightarrow{\text{meiosis}}$ 4 cells of 32 chrom. each $\xrightarrow{\text{meiosis}}$ 16 cells of 16 chrom. each $\xrightarrow{\text{meiosis}}$ 64 cells of 8 chrom. each

Clearly this process would be ineffective for the growth and maintenance of an organism.

However, the life cycles of both plants and animals indicate that meiosis is very important to sexual reproduction. If the fusion of gametes from two individuals is to result in a zygote with the same number and same kind of chromosomes as the parents, meiosis must occur in the production of gametes. If meiosis did not occur, each gamete would have the same number of chromosomes as the parent ($2N$), and the resulting zygote would have twice that number ($4N$). The meiotic process produces gametes with only half the parental chromosomes, one from each homologous pair. By this method a constant number of paired chromosomes is maintained in each species which reproduces sexually.

Suggested Reading

MAZIA, DANIEL. "How Cells Divide," in *The Living Cell*, W. H. Freeman, San Francisco, Calif. 1965.

11 An Introduction to Genetics

Genetics is the study of heredity. It investigates both in transmission of hereditary traits within families or populations and the expression of hereditary traits within the individual. In the mid-nineteenth century, Gregor Mendel studied hereditary patterns by performing many breeding experiments with plants in a monastery garden. From a statistical analysis of the results of these breeding experiments, he arrived at a series of important conclusions about the mechanisms of heredity without knowing a thing about the hereditary material of the cell. It is known now that the cellular basis for the principles he discovered rests in the characteristics and behavior of chromosomes during meiosis.

Mendelian Genetics

In a classic experiment, Mendel crossbred tall garden pea plants with dwarf garden pea plants. The ancestors of the tall plants had all been tall, and those of the dwarf plants had all been dwarfs. All of the offspring of these crosses were tall. He then crossed the offspring to one another and found, with a sizeable sampling, that only 3/4 of the offspring of these crosses were tall. That is, 1/4 of all the offspring of these tall parents were dwarf plants. This cross is schematically represented in Table 11-1.

On the basis of these and similar experiments, Mendel was able to make some generalizations about heredity. It is important to remember that Mendel's conclusions are based on the distribution of traits among the offspring from a large number of *identical crosses*.

The observed percentage of offspring with a certain trait from a large sampling of identical crosses represents the statistical probability that the particular trait will result from any specific cross in that sampling. For example, the fact that a population of humans is approximately half male and half female indicates that any individual couple has a 50% chance of having a boy and a 50% chance of having a girl.

Since he observed a significant number of offspring in the F_2 with a trait that had disappeared in the F_1, Mendel postulated the *law of segregation*. This law proposed that the hereditary information governing a particular trait is carried in a particulate factor (now called a gene), which does not disappear but remains stable from generation to generation. Further, these genes occur in pairs in an individual, and the effects of one may "mask" the effects of the other. Then, in the formation of gametes these genes segregate so that each gamete receives only one gene governing a particular trait.

This combination of assumptions comprising the law of segregation can explain Mendel's experimental results. If both the genes governing height in the tall P_1 plant are for tall (TT), and both the genes for height in the dwarf P_2 plant are for dwarf (tt), then the gametes of P_1 will all have a T gene and those of P_2 will all have a t gene. The F_1 plants, then, will all have a Tt gene pair.

The combination of genes that an organism has for a particular trait or traits is called its *genotype*. The genotype for all F_1 plants therefore is Tt. If the expression of the tall character T "masks" that for the dwarf t, all of these Tt plants will be tall. The tall

Table 11-1
Cross of Tall and Dwarf Garden Pea Plants

Explanation	Genetic notation	Characteristics
Original cross (parents with different traits)	$P_1 \times P_2$	tall × dwarf
Offspring (first filial generation)	F_1	all tall
Second cross	$F_1 \times F_1$	tall × tall
Offspring (second filial generation)	F_2	3/4 tall, 1/4 dwarf

character T is considered *dominant* and the dwarf character t, *recessive*. In genetic notation, a dominant gene is capitalized and a recessive gene is not. The *phenotype* is the appearance or behavior of the organism that results from its genotype. Phenotypically, the F_1 plant is tall, even though it carries a dwarf gene.

The F_1 plant produces two kinds of gametes in equal proportions, one with a T gene and the other with a t gene. When these gametes fuse randomly, any egg can combine with any sperm. The zygotes from the T eggs will be half TT and half Tt. Similarly, all t eggs will give half Tt and half tt zygotes. Since half of the eggs are T and the other half t, the genotypes TT, Tt, and tt will be formed in a 1:2:1 ratio. The phenotypes will be 3/4 tall and 1/4 dwarf since three out of four zygotes have a dominant T gene.

Mendel deduced the existence of genes and their behavior during gamete formation without knowing anything of their cellular or chemical nature. Since chromosomes are now known to contain the hereditary material, the gene behavior described by Mendel is readily explained by the nature of chromosomes and their behavior in meiosis. The pair of "particulate factors" governing a particular trait which Mendel postulated are alleles on a homologous pair of chromosomes. The homologous chromosome pair carrying the genes for height of Mendel's sweet pea plant is represented diagramatically in Fig. 11-1. Both alleles of the P_1 plant must be T and those of the P_2 plant are t. If the two alleles governing a particular trait in an organism are identical, the organism is *homozygous* for that gene. If the alleles differ, the organism is *heterozygous*. During anaphase I of meiosis, the homologous chromosomes move to the opposite poles of the cell. The segregation of genes that Mendel hypothesized occurs, then, at this stage. The homozygous tall plant forms only gametes with T genes and the homozygous dwarf plant forms only gametes with t genes. The F_1 resulting from the fusion of these two different gametes will be heterozygous Tt.

When the heterozygous homologous chromosome pair of the F_1 separates in meiosis, however, two kinds of gametes are formed in equal numbers (Fig. 11-2, top). One carries the T allele and the other the t allele. The distribution of the F_2 genotypes which results from random fusion of these F_1 gametes can be determined with a Punnett square diagram. A Punnett square diagram (Fig. 11-2, bottom) is a convenient method for determining all the possible zygote combinations that can result from a particular cross. In this checkerboard diagram every possible egg is paired with all possible sperm. Of the four possible combinations shown, three have the dominant T gene, so 3/4 of the offspring are tall and 1/4 dwarf.

Mendel also crossbred plants that differed in two traits, using

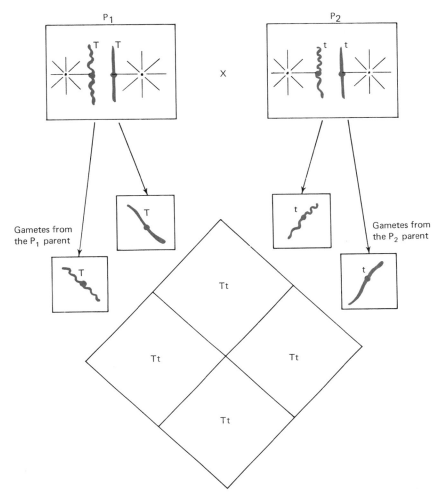

11-1 A diagramatic illustration of gamete formation resulting from meiosis in the tall pea plants (P₁) and dwarf plants (P₂). Random combinations of any P₁ gamete with any P₂ gamete all result in a heterozygous Tt genotype for the F₁.

parents that were homozygous for both traits. One such cross was between a plant which produced the dominant round (R) and yellow (Y) seeds and one that produced the recessive wrinkled (r) and green (y) ones. Again, all the F_1 plants produced the dominant round, yellow seeds. The F_2, however, gave the following ratio of phenotypes:

9/16 round and yellow seeds
3/16 round and green seeds
3/16 wrinkled and yellow seeds
1/16 wrinkled and green seeds

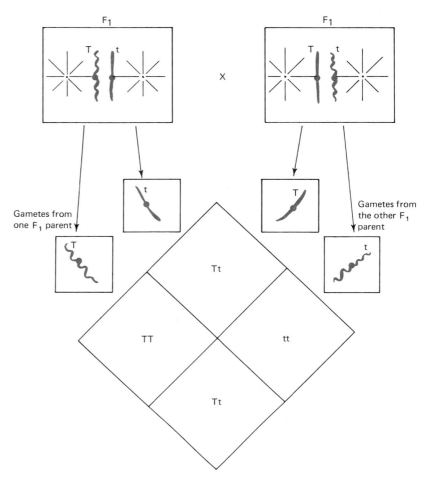

11-2 A diagramatic illustration of gamete formation resulting from meiosis in F_1 pea plants. Each plant produces two different gametes. A Punnett square conveniently pairs both gametes of one F_1 parent with both gametes of the other and shows the possible resulting F_2 zygotes.

These results led him to propose the *law of independent assortment*. This law states that the alternant genes of one trait separate at gamete formation independently from those of another trait.

This law explained fully his experimental results. Since both parents were homozygous in both traits, each could only form one type of gamete. All the gametes of the dominant plant carry both the R and the Y genes. Similarly, all the gametes of the recessive plant carry both the r and the y genes. The F_1 plants resulting from fusion of these two different gametes will be heterozygous RrYy. If the genes separate independently, the F_2 offspring will have the distribution of an Rr × Rr cross and of a Yy × Yy cross. That is, seed shape will be independent of seed color.

For a Rr × Rr cross, 3/4 of the offspring plants will produce round seeds and 1/4 will produce wrinkled seeds. Similarly, for a Yy × Yy cross, 3/4 of the offspring will give yellow seeds and 1/4 will have green seeds. Therefore, of the plants producing yellow seeds, 3/4 of them will also give round seeds and 1/4 will give wrinkled seeds. Only 3/4 of all the plants give yellow seeds, so 9/16 (3/4 × 3/4) will give round, yellow seeds, and 3/16 (3/4 × 1/4) will give wrinkled, yellow seeds. Similarly, of the green seed producing plants, 3/4 will also give round seeds and 1/4 will give wrinkled seeds. Since only 1/4 of all plants give green seeds, 3/16 (1/4 × 3/4) will give green and round seeds and only 1/16 (1/4 × 1/4) will give green and wrinkled seeds. This distribution of phenotypes is precisely that which Mendel observed.

At the cellular level the law of independent assortment results from the independent and random alignment of all chromosome pairs in metaphase I of meiosis. The two possible chromosomal alignments for a cell with the F_1 genotype RrYy are schematically displayed in Fig. 11-3. These two different alignments give rise to four different gametes, RY, Ry, rY, and ry. Since the two alignments

11-3 A diagramatic illustration to show that the random alignment of chromosomes during metaphase I of meiosis results in the independent assortment of genes among the gametes.

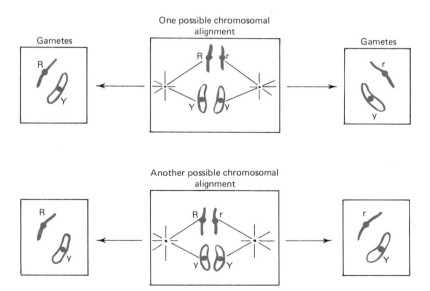

11-4

The Punnett square method for obtaining all of the random combinations of gametes from parents during sexual reproduction. This table presents all the possible zygotes and the frequency with which they occur.

	RY	Ry	rY	ry
RY	RRYY	RRYy	RrYY	RrYy
Ry	RRYy	RRyy	RrYy	Rryy
rY	RrYY	RrYy	rrYY	rrYy
ry	RrYy	Rryy	rrYy	rryy

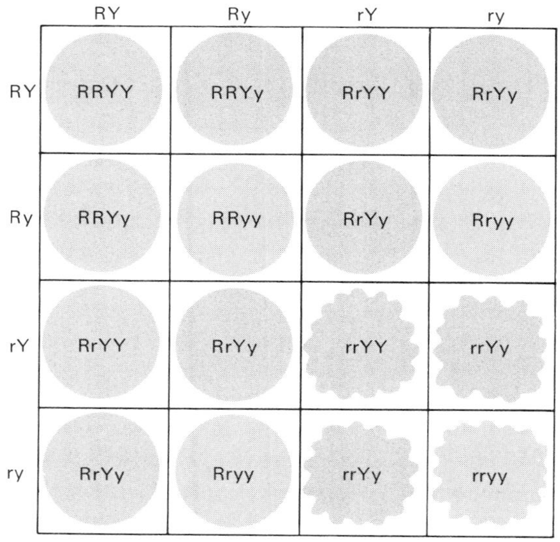

are equally probable at metaphase I, the different gametes are all equally probable.

In a cross between the F_1 plants these gametes combine randomly. A Punnett square is used to illustrate the possible zygote combinations from the F_1 cross and the frequency with which they occur (Fig. 11-4). This square shows that of the 16 possible genotype combinations, nine combinations contain both the dominants R and Y (R_Y_). There are six combinations which have a dominant gene for one trait but are homozygous recessive in the other, three of genotype R_yy and three of genotype rrY_. There is only one combination that is homozygous recessive for both traits (rryy). There are then four phenotypes displayed but nine genotypes as shown in Table 11-2.

Table 11-2

Genotypes and Phenotypes Showing Independent Assortment

Genotype	Frequency	Phenotype
RRYY	1 ⎫	
RrYY	2 ⎬ 9/16	round, yellow
RRYy	2 ⎪	
RrYy	4 ⎭	
RRyy	1 ⎫ 3/16	round, green
Rryy	2 ⎭	
rrYY	1 ⎫ 3/16	wrinkled, yellow
rrYy	2 ⎭	
rryy	1 ⎬ 1/16	wrinkled, green

Clearly the random distribution of homologous chromosomes during gamete formation and the subsequent random combination of gametes explain fully the process of independent assortment Mendel discovered in 1860.

Lack of Dominance

In the cases of competing genetic characteristics that Mendel studied (tall vs. dwarf, round vs. wrinkled seeds, etc.), one trait exhibited complete dominance over the other. This situation made his results easier to interpret. There are many cases in the biological world, however, where one allele does not totally "mask" the expression of the other. The genetic determination of blood type in humans is an example exhibiting this lack of dominance. There are three alleles governing blood type, commonly referred to as A, B, and O. Both A and B are dominant over O. Hence genotypes AA and AO exhibit phenotype A and genotypes BB and BO give rise to phenotype B. O is recessive and the phenotype O only occurs if the genotype is homozygous OO. The alleles A and B, however, are not dominant over one another. Individuals with an AB genotype do not have either an A or a B phenotype, but a different phenotype, AB. Evidently both alleles are expressed, neither dominates or masks the expression of the other.

Gene Linkage and Genetic Crossover

Mendel's principle of independent assortment holds only, of course, for genes on different (nonhomologous) chromosomes. Each chromosome, however, consists of many genes linked together, and these linked genes will not separate during meiosis unless chromosomal crossover occurs. An example of the genetic difference between linked and nonlinked genes is illustrated in Fig. 11-5. If an organism is heterozygous for three nonlinked traits, AaBbCc, it can form eight different gametes. If the genes are linked, only two different gametes are possible if crossover does not occur.

During prophase I of meiosis, the chromatids of homologous chromosomes coil around each other and sometimes come into contact. The point of contact is often visible under a microscope as an X-shaped figure called a chiasma. At this point the chromatids may exchange parts, so that the pieces of the chromatids in contact become attached to the chromatids of the homologous chromosome rather than the chromatids from which they came (Fig. 11-6). This process is referred to as genetic crossover.

The formation of new gene combinations of nonallelic linked genes depends on a chance chromosomal contact at some point along the chromosome between the nonallelic genes. The crossover probability is greater, therefore, the farther apart the two linked genes are on a chromosome. Linked genes that are far enough apart can recombine as frequently as genes on nonhomologous chromosomes. Genetic crossover makes possible, in theory at least, all combinations of nonallelic genes on the same chromosome.

The phenomenon of crossover has been useful in determining the location and arrangement of genes along a chromosome. This process is called chromosome mapping. The frequency at which crossover occurs between two genes has been shown to be proportional to the distance between them. The distance between genes

11-5 A diagramatic comparison of the number of combinations possible with three heterozygous gene pairs when they are on separate chromosomes (top) and linked (bottom).

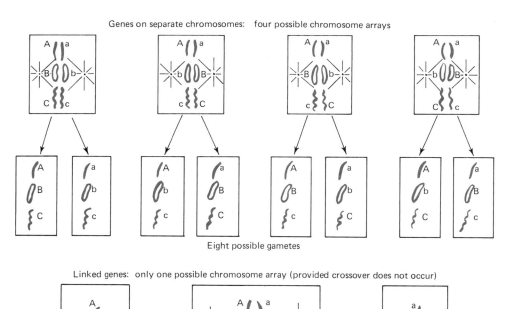

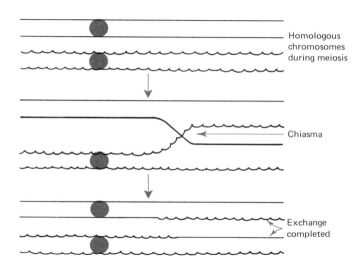

11-6

A diagramatic presentation of genetic crossover. The homologous maternal and paternal chromosomes pair during meiosis (top), cross chromatids at a chiasma (middle), and exchange parts (bottom).

is set equal to the percent of the offspring which results from a crossover occurring between them.

The process of chromosome mapping is best illustrated by an example. Consider a hypothetical cross of a parent homozygous for the linked dominant genes A, B, and C, with a parent homozygous for the recessive alleles a, b, and c. The F_1 would be heterozygous for all characters. During meiosis the F_1 should form only two types of gametes, all recessive abc, or all dominant ABC. If the F_1 is crossed with a homozygous recessive individual, it is possible to determine from the offspring phenotypes if crossover occurred during F_1 meiosis. Offspring in which the three traits are not all dominant or all recessive can only result from crossover. The results of such a cross are shown below:

$$\frac{ABC}{abc} \times \frac{abc}{abc}$$

Out of 1282 offspring:

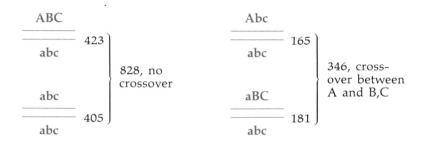

$$\underline{\dfrac{ABc}{abc}} \Biggr\} \begin{array}{l} 5 \\ \\ \\ 7 \end{array}$$

$$\underline{\dfrac{abC}{abc}}$$

12, crossover between C and A,B

$$\underline{\dfrac{AbC}{abc}} \Biggr\} \begin{array}{l} 53 \\ \\ \\ 43 \end{array}$$

$$\underline{\dfrac{aBc}{abc}}$$

96, crossover between B and A,C

The frequency of crossovers between A and B is

$$346 + 96 = 442 \qquad 442/1282 = 0.35$$

Crossover between genes A and C occurs at a frequency of

$$346 + 12 + 358 \qquad 358/1282 = 0.28$$

The frequency of crossovers between genes B and C is

$$96 + 12 = 108 \qquad 108/1282 = 0.08$$

The map distances in map units (mu) between each gene pair is equal to the percent offspring resulting from crossover between those two genes. In this example the map distances are

Between A and B = 35 mu
Between A and C = 28 mu
Between C and B = 8 mu

Since genes A and B are the farthest apart, they must be the two terminal genes of the trio. The chromosome map becomes:

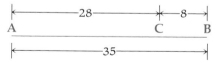

In this calculation the distance between genes A and B is similar to, but not equal to, the sum of the distances between A and C plus C and B. This inequality occurs because the 12 offspring (ABc and abC phenotypes) resulting from crossover between C and A,B must arise from a double crossover as indicated below:

$$\underline{\dfrac{A \quad C \quad B}{a \quad c \quad b}} \longrightarrow \underline{\dfrac{A \quad c \quad B}{a \quad C \quad b}} \longrightarrow \underline{\dfrac{A \quad c \quad B}{a \quad C \quad b}}$$

Crossover between genes A and B actually occurs twice to form these offspring. These individuals, however, were not included in the calculations of crossovers between A and B since they do not represent recombinations between genes A and B. Map distances between outside markers only equal the sums of the inside distances when there are no double crossovers.

This example illustrates how useful chromosome mapping is in locating genes on chromosomes.

Practice Exercises in Genetics

Some genetics problems follow which can be answered by employing the genetic principles presented in this chapter. The answers are given at the end of the problem set.

1. SAMPLE PROBLEM: The gene for tall is dominant over dwarf in the garden pea plant used by Mendel. A pea plant that comes from a line of plants that are all tall is crossed with a dwarf pea plant. What is the phenotype of the F_1 generation? What is (are) its genotype(s)?

2. If the offspring generation of problem 1 is crossed with the tall plant from a tall lineage, what will be the phenotype(s) and in what ratios for the offspring? What will be the genotype(s) and in what ratios?

3. If the F_1 generation of problem 1 is crossed with the dwarf parent from a dwarf lineage, what will be the genotype(s) and the ratios of the offspring, and the phenotype(s) and ratios of the offspring?

4. The genes for dark eyes (black or brown) are usually dominant over genes for blue or gray eyes. A man with black eyes marries a woman with light gray eyes. They have two children, a boy with black eyes, and a girl with blue eyes. What are the genotypes of the man, his wife, the little boy, and the little girl?

5. A man with brown eyes marries a woman with blue eyes. They have 12 brown-eyed children. What are the genotypes of the man, his wife and all the children?

6. A blue-eyed man marries a blue-eyed woman. What is the probability of their having a brown-eyed child.

7. A brown-eyed man marries a blue-eyed woman. They have four children, two with brown eyes, and two with blue eyes. What are the genotypes of all these people?

8 A black-eyed man marries a black-eyed woman. They have a gray-eyed child. What are the genotypes of the father, mother, and child? The probability of their next child having gray eyes is what?

9 A dark-eyed man has a dark-eyed son, who marries a gray-eyed woman. What is the maximum probability of their having a dark-eyed child? Why?

10 Assume that the dimple is inherited as a simple dominant gene. A dimpled man whose mother had no dimple marries a woman with no dimple. What is the probability that they will have a child with a dimple?

11 A brown-eyed man with a blue-eyed mother marries a brown-eyed woman with a blue-eyed father. What is the probability that their first child will be brown-eyed? That the second child will be brown-eyed?

12 A man and woman have 24 children. Of the children, 17 have brown eyes and 7 of the children have blue eyes. What are the genotypes of the parents?

13 Assume that blood type is inherited as A and B dominant over O, but A and B incompletely dominant over each other. Genotypes AA and AO are then phenotypically type A, genotypes BB and BO are type B, genotype AB is type AB, and genotype OO is type O blood. A man with type A blood marries a woman with type A blood. They have the first child as blood type O. What are the genotypes of the father, mother, and baby?

14 A man with type AB blood marries a woman with type O blood, but whose father was type A blood. What genotype would you expect their first child to have?

15 A man with type B blood marries a woman with type A blood. They have six type AB children. What are the genotypes of the father, mother, and children?

16 A man with type A blood, but a type O father, marries a woman with type O blood. What blood types would the children have, and in what theoretical proportions?

17 A man whose father is type B and whose mother is type A, has blood type A. He marries a type A woman, whose parents had the same blood types as his parents. What are the geno-

types of the man and woman, and what is the probability that their first child will be blood type A?

18 A type A man whose mother was type O marries a woman with type B blood. Their son has type B blood. This son marries a girl with type B blood. They have 12 children, 10 type B and 2 type O. What are the genotypes of the man, woman, son, girl, and children?

19 A man with type AB blood marries a woman with type A blood. They have a type B child. What are the genotypes of the father, mother, and child?

20 A man with type A blood marries a woman with type B blood. They have a type O child. What is the probability of their fifteenth child having type O blood?

21 A man with type AB blood marries a woman with type AB blood. What is the probability of their first child being type O?

22 A man whose father was AB and whose mother was B, has type A blood. He marries a woman with type A blood but whose father was type A, but whose mother was type B. What is the probability that the first child will be type A? What is the probability that the second child will be type A? What is the probability that the third child will be type A?

23 A man with type A blood marries a woman with type A blood. They have eight type A children, one type O child. What are the genotypes of the father, mother, the eight type A children, and the one type O child?

24 A brown-eyed man marries three times. His first wife is blue-eyed. They have two children, one brown-eyed and one blue-eyed. His second wife is brown-eyed. They have two children, both blue-eyed. His third wife is brown-eyed. They have six children, all brown-eyed. What are the genotypes of all the people involved?

25 A man with type A blood marries three times. His first wife is type B. They have three children, types AB, A, and A. This man marries again, this time to a woman with type A blood. They have two children, both type A. This man marries a type O woman, and they have four children, all type A. What are the genotypes of all these people?

An Introduction to Genetics

26 (*Note:* These are dihybrid crosses; refer to the diagram on page 154 of the text for an example problem.) A man with a dimple and brown eyes (whose father had blue eyes but no dimple) marries a woman with a dimple and brown eyes (whose father had blue eyes with no dimple). What is the probability their first child will be blue-eyed and without a dimple? (Assume that dimple is a dominant over smooth cheeks, and brown eyes are dominant over blue.)

27 A blue-eyed woman with no dimple marries a man with brown eyes and a dimple (whose mother had blue eyes and no dimple). What is the probability that their first child will have blue eyes and a dimple?

28 A man with blue eyes and a dimple (whose mother had blue eyes and no dimple) marries a woman with blue eyes and no dimple. What is the probability that their first child will be blue-eyed with no dimple?

29 A woman with type AB blood and brown eyes marries a man with type A blood and brown eyes. Their first child has type B blood and blue eyes. What are the genotypes of the parents and their child?

30 The two children of a family have genotypes OOdd and BBDD for blood type and dimple. What are the genotypes of the parents?

31 A fruit fly with genes CDE homozygous dominant is crossed with a fruit fly homozygous recessive for these genes, giving the genotype CcDdEe. These genes are carried on one pair of homologous chromosomes. This F_1 generation is crossed with fruit flies homozygous recessive for these genes. The genotypes of the resulting offspring, and their numbers, are given below (only expressed genotypes are given, since only the combinations of dominant and recessive genes are used in calculating the answer):

CDE	432	CdE	22
cde	451	cDe	27
CDe	44	Cde	2
cdE	60	cDE	2

What is the order of these genes on the chromosome, and what are the approximate distances between them on the chromosome? (See pages 157 and 158 for sample solution.)

32 A fruit fly with genes ABC homozygous dominant is crossed with a fruit fly homozygous recessive for these genes, giving the genotype AaBbCc in the F_1 generation. These genes are carried on one pair of homologous chromosomes. This F_1 generation is crossed with fruit flies homozygous recessive for these genes. The genotypes of the resulting offspring and their numbers, are given below (only expressed genotypes are given since only the combinations of dominant and recessive genes are used in calculating the answer):

ABC	454	ABc	25
abc	497	abC	38
AbC	95	Abc	2
aBc	82	aBC	6

What is the order of the genes on the chromosome, and what are the approximate distances between the genes on the chromosome?

33 A fruit fly with genes XYZ homozygous dominant is crossed with a fruit fly homozygous recessive for these genes, giving the genotype XxYyZz in the F_1 generation. These genes are carried on one pair of homologous chromosomes. This F_1 generation is crossed with fruit flies homozygous recessive for these genes. The genotypes of the resulting offspring, and their numbers, are given below (only expressed genotypes are given since only the combinations of dominant and recessive genes are used in calculating the answer):

XYZ	797	Xyz	84
xyz	819	xYZ	80
XYz	120	XyZ	6
xyZ	116	xYz	8

What is the order of these genes on the chromosome, and what are the approximate distances between these genes on the chromosome?

34 A fruit fly with genes RST homozygous dominant is crossed with a fruit fly homozygous recessive for these genes, giving the genotype RrSsTt in the F_1 generation. These genes are carried on one pair of homologous chromosomes. This F_1 generation is crossed with fruit flies homozygous recessive for these genes. The genotypes of the resulting offspring and their numbers, are given below (only expressed genotypes are given

An Introduction to Genetics

since only the combinations of dominant and recessive genes are used in calculating the answer):

RST	629	rsT	93
rst	606	rSt	3
RsT	3	Rst	35
RSt	88	rST	41

What is the order of these genes on the chromosome, and what are the approximate distances between these genes on the chromosome?

35 A fruit fly with genes EFG homozygous dominant is crossed with a fruit fly homozygous recessive for these genes, giving the genotype EeFfGg in the F_1 generation. These genes are carried on one pair of homologous chromosomes. This F_1 generation is crossed with fruit flies homozygous recessive for these genes. The genotypes of the resulting offspring and their numbers, are given below (only expressed genotypes are given since only the combinations of dominant and recessive genes are used in calculating the answer):

EFG	369	eFG	50
efg	375	efG	74
Efg	43	EfG	4
EFg	71	eFg	3

What is the order of the genes on the chromosome, and what are the approximate distances between the genes on the chromosome?

Answers to Practice Exercises

1 Tall; Tt.

2 All tall; 1/2 TT; and 1/2 Tt.

3 1/2 tall; 1/2 dwarf; 1/2 Tt and 1/2 tt.

4 Man Dd; woman dd; boy Dd; girl dd.

5 Man DD; woman dd; all children Dd.

6 Small.

7 Man Bb; woman bb; 2 children Bb; 2 children bb.

8   Man Dd; woman Dd; child dd; Dd × Dd gives 1/4 dd.

9   Maximum 100%; minimum 50%. Depends on whether father is Dd or DD; DD would give 100%, whereas Dd would give 50%.

10 1/2.

11 3/4; 3/4; Dd × Dd gives 3/4 DD or Dd.

12 Man probably Bb; woman probably Bb. Their children lie close to the 3/4–1/4 ratio of brown eyes versus blue eyes expected of the Bb × Bb cross.

13 Man AO; woman AO; child OO.

14 1/2 for AO; 1/2 for BO.

15 Man probably BB; woman probably AA; children all AB. Chances very small of getting six consecutive AB children if either parent is heterozygous.

16 1/2 type A (genotype AO); and 1/2 type O (genotype OO).

17 Father AO; mother AO; 3/4.

18 Man AO; woman BB or BO; son BO; girl BO; children 10 BB or BO, 2 OO.

19 Man AB; woman AO; child BO.

20 1/4 for type O.

21 None.

22 3/4; 3/4; 3/4.

23 Man AO; woman AO; 8 children AA or AO, 1 child OO. In this case the OO did not occur in its theoretical 1/4 frequency.

24 Man Bb; first wife bb; children Bb and bb.

   Second wife Bb; children bb and bb.

   Third wife probably BB; all children either BB or Bb.

25 Man AA; first wife BO; children AB, AO, AO.

   Second wife AA or AO; children AA or AO.

   Third wife OO; children all AO.

26 1/16.

27 1/4.

28 1/2.

29 Man A/OD/d; woman A/BD/d; first child B/Od/d.

30 Both parents have to be B/OD/d.

31 DCE or ECD

32 CAB or BAC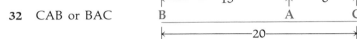

33 XYZ or ZYX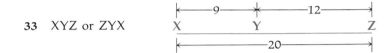

34 RST or TSR

```
|←—5—→|←————12————→|
R     S            T
|←—————17—————→|
```

35 EFG or GFE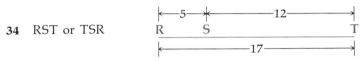

Part IV

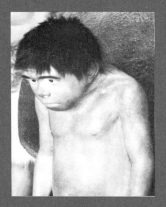

Evolution

Restoration of Neanderthal man. (Courtesy of Field Museum of Natural History, Frederick Blaschke sculpture, Chicago, Illinois.)

The Role of the Organism in Evolution 12

Broadly speaking, evolution is the process by which the organisms on earth today arose from nonliving matter. The theory of the evolution of primitive forms of life from the chemicals composing the primordial earth was presented in detail in Chapter 1. The theory of chemical evolution is based on current knowledge of the nature of matter and conjectures as to the conditions on the earth at the time. The theory as to how the myriads of organisms on earth today arose from primitive life forms is based on current knowledge of how cells and organisms behave, reproduce, and interact with their environments and conjectures as to the prevailing environments during the various stages of the development of the earth.

In 1859 Charles Darwin proposed a theory for biological evolution which is still valid today. Darwin's theory is based on the conclusions he drew from three observations he made about organisms and their populations. He observed that the *reproductive potential* of organisms exceeds their possible survival rate. That is, organisms can produce greater numbers of offspring than can survive in their environment. He noticed, too, that populations remain fairly constant in size. His third observation was that there is considerable *variability* within a species. Any population of a given organism exhibits a wide spectrum of genetic and phenotypic variations. From these observations Darwin concluded both that all organisms cannot survive and that, since organisms differ from one another, then the organisms which do survive, on the whole, are the ones that are best suited to the environment. This is the essence of natural selection.

This chapter will be concerned with the reproductive potentials of organisms and the sources of variability within a population of organisms. These two factors dictate the role of the organism in evolution. The next chapter will discuss natural selection and the role of the environment in evolution.

Reproductive Potential of Populations

The carrying capacity of the environment is the number of organisms of each particular kind that an environment can support for a given time. If an organism were reproduced at a rate that does not exceed the environment's carrying capacity, the population of that organism would be stable. However, no existing species has ever exhibited this reproductive pattern. Instead, the reproductive rate of a species is so high that its population soon exceeds the carrying capacity of its environment. As a result, organisms must compete to survive.

An example of the reproductive capacity of a common housefly will serve as an illustration of this phenomenon. Assume that the female lays 100 eggs and that the sex ratio of the offspring is 50:50 male to female. The generation time for the housefly is about 30 days and the lifetime is less than 60 days. From these assumptions the theoretical fly population resulting from two young flies after six months is worked out in Table 12-1.

Table 12-1 Population Increase Table for the House Fly

Assume an insect population that starts with one male and one female.
Females lay 100 eggs; death after 30 days; 50:50 sex ratio.

Generation	Age group		Total Population
	Offspring	30 days	
1	1 + 1 = 2		2
2	50 + 50 = 100	1 + 1 = 2	102
3	2500 + 2500 = 5000	50 + 50 = 100	5100
4	125,000 + 125,000 = 250,000	2500 + 2500 = 5000	255,000
5	6,250,000 + 6,250,000 = 12,500,000	125,000 + 125,000 = 250,000	12,750,000
6	312,500,000 + 312,500,000 = 625,000,000	6,250,000 + 6,250,000 = 12,500,000	637,500,000

After 30 days these two original flies produce 100 offspring. In another 30 days the original pair will be dead but the 50 female offspring will each lay 100 eggs. The population, then, is up to 5100 in 2 months. If this pattern continues for three more generations, the number of houseflies can potentially go from 2 to 637,500,000 in 6 months. This could and would happen if the population was not controlled in any way and the food supply was unlimited.

In other chapters the potentials for population increase for the rabbit (Table 16-1), deer (Table 17-1), and man (Table 18-1) are presented. After five generations the theoretical population increases for these animals are from 1 to 80 for the rabbit, from 1 to 14 for the deer and from 1 to 30 for the human. Since the generation time is longer and the number of offspring is smaller for each of these organisms than they are for the fly, these potentials are not quite as overwhelming. They clearly show, however, the explosive potential for population increase inherent in the ability of an organism to reproduce itself.

Population increases of these magnitudes cannot continue indefinitely without the population exceeding the carrying capacity of the environment. When this happens, not all individuals in the population can survive. Clearly the ability of organisms to overproduce is an important factor in selecting out the best adapted individuals within a population as well as the best adapted population within a biological community.

Variability of Organisms

Variability amongst organisms arises from even small differences in the genetic material. Mutation is one mode of causing differences in genetic material. Chromosomal mutations, as discussed in Chapter 9, change the arrangement of genes on a chromosome by translocation, duplication, inversion, or deletion. Translocation and inversion are mechanisms by which the genes are rearranged on the chromosomes; duplication and deletion lead to the addition and loss, respectively, of chromosomal material. Although these mechanisms do not change the overall base content of DNA, they can alter the base sequences and, therefore, can produce great variation in the resulting organisms.

Mutations

Point mutation is a change in the genetic information resulting from a specific chemical change of the base content within the gene. A point mutation, then, actually changes a gene; chromosomal mutation merely rearranges it. Both processes necessarily play a large part in introducing variability amongst organisms and, hence, in the overall evolutionary process.

The rate of mutation is relatively slow. That is, it is not a probable occurrence during most reproductions. However, that it does occur and does so at a slow rate makes it an effective mechanism in evolution. If mutation did not occur, there would be no change and little, if any, variation in organisms. On the other hand, if it were a much more rapid process, organisms would change constantly from generation to generation. Even the organisms which were well adapted to their environment would disappear. The rate of spontaneous mutation leads to a relatively stable population which has a potential for small changes to appear within it. These changes may or may not cause some individuals in the population to be better adapted to their environment.

An example might best illustrate this point. Clearly there are visible differences between a collie, a beagle, a labrador, a setter, a spaniel, and so forth, yet they are all dogs and can all interbreed. In fact, the genetic differences between them are less than 1% of their genetic make-up. Small changes, then, have occurred in the genetic character of the dog population. The basic characteristics of dogs, however, remain stable. These small variations, however, have produced breeds. In this example, the variations of each breed have been artificially selected for by dog breeders in a relatively stable population of dogs.

Genetic Recombination by Meiosis-Fusion

Sexual reproduction is a major source for variability within organisms as it generally leads to offspring with a different combination of genes than those of either parent or any other ancestor. During meiosis, the homologous chromosome pairs segregate randomly and independently of one another. Most of the resulting gametes receive genes from both parents. Zygote formation is also a random process. Any gamete from one individual has an equal chance of fusing with any gamete of any other individual within the population. These two random processes inherent in sexual reproduction introduce a great deal of variability within a population.

Because of the mechanism of sexual reproduction, two individuals can give birth to offspring which are better or less suited to the environment than they were themselves. An example best illustrates this point.

Suppose gene G on one chromosome is dominant and controls a trait that is advantageous to the organism. Another dominant

		GB	Gb	gB	gb
	GB	G GB B	G GB b	G gB B	G gB b
Parent II	Gb	G GB b	G Gb b	G gB b	G gb b
Genotype G gB b	gB	G gB B	G gB b	g gB B	g gB b
	gb	G gB b	G gb b	g gB b	g gb b

Parent I
Genotype
G gB b
Gametes

12-1

An example of gene recombination from meiosis-fusion. In this cross G and b are the most advantageous genes. Neither parent has the most favorable genotype, but 3/16 of the progeny does; 3/16 of the progeny also has the most unfavorable genotype, ggB__.

gene, B, on a different chromosome, however, is disadvantageous. The homozygous recessive allelic condition, bb, though, leads to an advantageous phenotype. The best adapted individuals then would have genotype G_bb with respect to these two genes. The Punnett square diagram in Fig. 12-1 shows the offspring of a cross between individuals with genotypes GgBb. Of their offspring, 3/16 will have the more advantageous genotype G_bb and will have a better chance of survival than their parents. In addition, 3/16 of their offspring will have the ggB_ genotype for both disadvantageous traits and will probably have a lesser chance of survival. In short, because of genetic recombinations during sexual reproduction, advantageous and deleterious genes are not lost from the population. They are lost only if they are selected against because the organisms that carry them die or do not reproduce for some reason.

Genetic Recombination by Crossover

Although genes of different chromosomes recombine by independent assortment during sexual reproduction, those on the same chromosome do not. Genes on the same chromosome only recombine if chromosomal crossover occurs during meiosis. Since crossover probability is higher the farther apart two linked genes are, the recombination frequency for linked genes increases as the distance between the two genes increases.

The variation in organisms due to this process may also lead to organisms that are better adapted to their environment. For instance, consider the example given earlier in which G is dominant and advantageous and b is recessive and advantageous only when homozygous. This time, however, assume these two genes are on

The Role of the Organism in Evolution   **[173]**

the same chromosome. Furthermore, in both GgBb parents the dominant genes are linked and the recessive genes are linked. Each parent can only produce GB and gb gametes. No crosses between these gametes would result in offspring with the advantageous genotype G_bb unless crossover occurs. Crossover could lead to a Gb gamete which, when crossed with a gb gamete, would produce an offspring with the advantageous G_bb genotype.

If the crossover frequency between the two genes is known, the percent of the offspring with the advantageous G_bb genotype can be calculated. If genes G and B are 10 mu apart, crossover occurs 10% of the time. The meiotic process in which this crossover occurs (for a GgBb individual) can be schematically represented as follows:

$$\frac{G\quad B}{g\quad b} \xrightarrow{crossover} 90\% \frac{G\quad B}{g\quad b} \quad 10\% \frac{G\quad b}{g\quad B}$$

$$\downarrow segregation$$

$$45\% \frac{G\quad B}{} \quad 45\% \frac{g\quad b}{} \quad 5\% \frac{G\quad b}{} \quad 5\% \frac{g\quad B}{}$$

The gamete frequencies from this parent are 0.45 GB, 0.45 gb, 0.05 Gb, and 0.05 gB. If this parent is crossed with a ggbb individual who has only gb gametes, the resulting genotypes of the offspring are 0.45 GgBb, 0.05 Ggbb, 0.45 ggbb, and 0.05 ggBb. That is, 1/20 of the offspring have the more advantageous genotype G_bb. Were it not for crossover, this genotype would not occur in this population. The frequency at which it occurs in the population is proportional to the crossover rate between the two genes.

Variability in Organisms and Populations

A population is a group of organisms which interbreed. In evolution studies it is more relevant to think in terms of the genetics of a population rather than of an individual. To this end it is convenient to employ the concept of a gene pool. All the genes from all the individuals of a population constitute the "gene pool" of that population. If the population is large enough and mating occurs randomly, the overall genetic make-up of the offspring of the population can be determined from the gene pool of the population, independent of the individual crosses.

For every trait there are a certain number of alleles which govern it. Each alternate allele occurs with a certain frequency within the gene pool. These alleles will be present in the same frequency in all the gametes (male and female) as they are in the gene pool. Since any gamete has an equal chance of meeting and fusing with any other, the genotypes of the progeny of a population can be determined statistically from the frequency of each allele in the gene pool. For example, if one allele B has a frequency q, then the other allele b has a frequency $(1 - q)$. The interbreeding in the population, then, is represented by the Punnett square below.

		Sperm gene frequencies	
		q B	$(1 - q)$ b
Egg gene frequencies	q B	q^2 BB	$q(1 - q)$ Bb
	$(1 - q)$ b	$q(1 - q)$ Bb	$(1 - q)^2$ bb

The progeny will be in the ratio:

$$q^2 \text{ BB}/2q(1 - q) \text{ Bb}/(1 - q)^2 \text{ bb}$$

The gene frequency in the next generation then becomes

For B: $\quad q^2 + q(1 - q) = q$
For b: $\quad q(1 - q) + (1 - q)^2 = q - q^2 + 1 - 2q + q^2$
$\quad\quad\quad\quad\quad\quad\quad\quad\quad\quad = 1 - q$

or the same as in the parental generation. This finding, that in the absence of outside forces the gene pool of a large randomly mating population is constant, is known as the *Hardy-Weinberg law*. It indicates that, in a population without mutation or natural selection in which random mating occurs, genes will not be lost, and all combinations of genotypes will continue to exist. That is, the population will be genotypically stable.

Genetic Drift

The Hardy-Weinberg law only holds if a population mates randomly in the absence of outside forces. Natural selection is the change in the gene pool due to those outside or environmental factors. Another deviation from the Hardy-Weinberg law occurs in small populations. If the population is very small, even though mating occurs randomly, the gene pool will not remain constant. Statistically, the gene frequencies of the small sample which perpetuates the population do not represent the gene frequencies of the original population. The process by which the gene pool of a small population changes is called genetic drift. The changes can be quite sharp and usually occur quite randomly. This point is

12-2

A simple example of genetic drift. A small population can get rapid shifts in gene frequency from the random frequency expected from chance combinations in breeding or survival due to the small number of combinations formed or surviving.

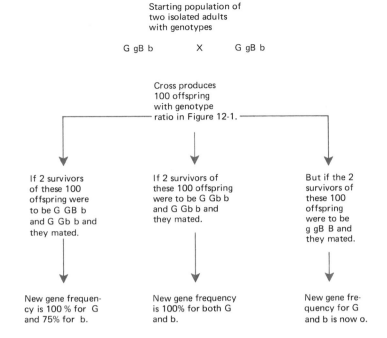

illustrated in the following small population. Consider a population of 100 insects resulting from a cross of GgBb × GgBb individuals where G_bb is the most advantageous genotype. Further, the mortality rate from generation to generation is 98%, so that only two individuals from the original cross perpetuate the population. In the original population the gene frequencies are 0.5 for each gene G, g, B, and b. There are nine different genotypes in the population, any one or two of which may perpetuate the insect population. As shown in Fig. 12-2, mating between the random survivors can drastically change the gene pool of the population. If the two survivors are GGBb and GGbb, the gene frequencies would change to 1.0 for G and 0.75 for b. If, however, the two survivors are both ggBB, the advantageous genes G and b would be lost from the population.

[176] Evolution

The Role of the Environment in Evolution 13

Every organism interacts with its environment. The environment is the sum total of all factors—physical, chemical, and biological—in an organism's surroundings with which it interacts. If its environment cannot supply all of its needs, an organism will die and/or be unable to reproduce. Since not all members of a population are alike, some of them will be better suited to the environment than others. In the long run, individuals of this type should have a better chance of survival than less well-adapted forms. Gradually then, the proportion of certain traits within the population will change. Natural selection is the process of differential reproduction of certain genotypes which cannot be attributed to chance alone. Because, as described in Chapter 12, organisms show a marked ability to overproduce and vary amongst themselves, natural selection is important in shaping the size and character of a population which a particular environment can maintain.

In order to study the changes that occur within a population due to its environment, it is convenient to use the species as a point of reference. A species is a population of organisms which share a common gene pool, that is, organisms that can only breed among themselves. The role of the environment in evolution can be seen by noting the changes in the gene pool of a species under different environmental situations: a stable environment, a changing environment, and an isolated environment.

Natural Selection in Stable Environments

If an environment is relatively stable for a long time, a species in that environment will gradually, through the generations, become better adapted to that environment. Some traits might appear within the species, either by mutation or genetic recombination, which make the possessors of that trait better suited to that environment. If these traits afford these individuals a better chance to survive and reproduce, the genes controlling them will, with time, occur in increasing frequency in the gene pool of the species. Once a species is well adapted to a stable environment, only small changes occur within the species (Fig. 13-1).

The less well-adapted forms of the species have much lower survival rates and hence may not reproduce. Their genes, therefore,

13-1 **The general process of selection against any deviations from the best-adapted form of an organism for a particular environment.** The deviations from the well-adapted form have less chance of survival than the well-adapted form.

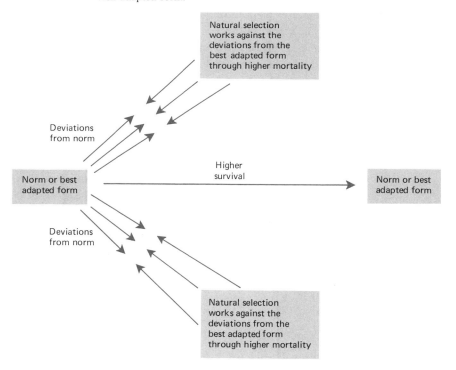

are slowly eliminated from the gene pool. For example, an albino deer is a poorly adapted form in any but a snowy environment, since he can be more readily spotted by predators than brown deer.

The basic feature of natural selection is differential reproduction. It is not sufficient for a better-adapted form merely to survive. It must also reproduce so that its advantageous genes get perpetuated throughout the species. Generally the better-adapted forms will survive at a higher rate to reproduce, or will be more successful in finding mates, or will obtain more food so that their young are more vigorous, and so on. Clearly, when differential reproduction occurs, the gene frequency of the advantageous genes will increase in the gene pool.

Effects of Environmental Change

In a stable environment, selection may be for a certain phenotype of a species. When the environment changes, a different phenotype may be selected for. An example of how a different dominant form of a species can result from a changing environment is found in the housefly population. Before World War II, most houseflies were sensitive to DDT. When DDT was introduced into the environment, most of the houseflies died. A few flies, however, were able to survive. These flies with DDT resistance represented only a small percent of the prewar housefly population. This phenotype had probably arisen by mutation but did not prove to be particularly advantageous in comparison to the phenotype of the DDT-sensitive fly. In the DDT environment, however, only these flies could survive. Perhaps when the competition of the possibly more robust DDT-sensitive fly was removed, any advantageous mutations that occurred in the DDT-resistant fly had a better chance for expression and perpetration because the DDT-resistant fly then had a better chance to survive. By this process a DDT-resistant fly, which was generally better suited to its environment, could develop.

Since the housefly has a high reproductive potential, these DDT-resistant flies were able to produce large numbers of DDT-resistant flies quickly in areas once dominated by the DDT-sensitive populations. This process, outlined in Fig. 13-2, has occurred with many insects and insecticides. In this example the environmental change is artificial or man-made, but the selection process which followed is the same as that resulting from natural environmental changes.

1,000,000,000,000,000,000
DDT sensitive houseflies

↓

DDT in environment

↓

A few DDT resistant houseflies survive

↓

Reproduction of the few DDT resistant houseflies in environment

↓

1,000,000,000,000,000,000
DDT resistant houseflies

13-2 A diagramatic presentation of the evolution of the DDT-resistant housefly. Intensive natural selection against DDT-sensitive houseflies when DDT came into use resulted in the selection for, and multiplication of, DDT-resistant houseflies.

The changed environment simply gives the genetic variations within a species a different chance at survival. A trait which adapted an organism well to the old environment may be unbeneficial or even deleterious to it in the new situation. Other traits found in a low percentage of the original population may be well suited to the new environment. Organisms, then, with this trait will tend to flourish. It is important to remember that the variations in the species exhibited in the different environments originate with the species and *not* with environment.

Isolation-Speciation

Random interbreeding within the species tends to distribute the genes of the species throughout the population. This process is called gene flow (Fig. 13-3a). Any advantageous spontaneous mutation occurring within the population would then be distributed through the population. If two interbreeding populations of a species are separated from one another, this gene flow is interrupted. The isolation can lead to differences between the two populations even though none originally existed. Theoretically, natural selection would maintain the same form of the organism in both populations provided the environments were identical and the same mutations occurred in each population. Both of these conditions are highly unlikely because environments are so complex that no two are exactly alike in all respects and because spontaneous mutations occur at random. Instead, for each population there is a different environment interacting with different mutations, and the two populations start to differ (Fig. 13-3b).

Ultimately, two such separated populations may differ enough that they are recognizably different, and they can no longer interbreed. At such time they become two new species (Fig. 13-3c).

There are several mechanisms by which populations are isolated from one another. Probably the most important is geographic isolation, in which a geographic barrier separates two populations for a long period of time. For example, in earlier times the Isthmus of Panama was under water. When the Isthmus of Panama emerged, it separated populations of many species of marine life between the Pacific Ocean and the Gulf of Mexico. On each side of this barrier these species changed in different ways because of the different forces they encountered until speciation occurred. This type of evolution has occurred over and over during geological history as oceans rose and receded, mountain ranges formed, and continents separated, and so on.

Ecological isolation also leads to speciation. Some species are ecologically unsuited to certain environments and cannot adapt to life in that environment. Hence, two populations of a species separated by such an ecologically unsuitable environment can no longer interbreed. These populations may speciate if their separate evolution is divergent enough.

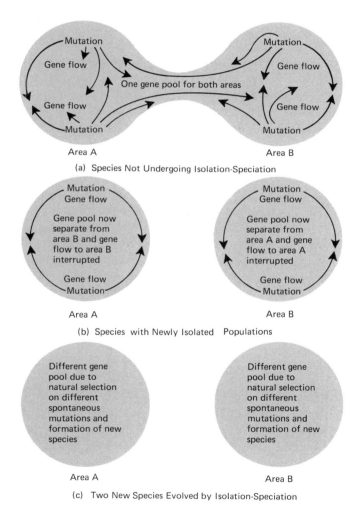

13-3

The basic mechanism of isolation-speciation. Two populations of one species become isolated from each other, and natural selection together with spontaneous mutations occurring in the isolated populations inevitably results in the formation of new species. When the new forms diverge to the point where they are recognizably different and no longer can interbreed, the process of speciation is accomplished.

The Role of the Environment in Evolution

Isolation mechanisms are not necessarily environmental, they can also be behavioral. That is, some populations which *can* interbreed may not do so because of differences in mating seasons or practices. If such exclusive behavior continues for a long period of time, so that each population changes through mutation and natural selection, speciation can result. The populations then are no longer *able* to interbreed.

Isolation, then, leads to divergence between two populations. It results from the same process of natural selection for the forms best adapted to an environment, which arises from the interaction of the environment and the variability of the organism.

Changing Gene Frequencies in Evolution

In effect, evolution is a change in the gene frequencies within the gene pool of a species. If a species is well adapted to its environment, this change will be slow. The gene pool will remain relatively stable unless a mutation arises which is even more advantageous to the species than its current dominant form.

Gene frequencies change if some forms of a species survive and reproduce differentially from other forms. If all forms that survive are equally fertile, a gene frequency change is due to differential survival. Death can occur within a population randomly or differentially. For instance, accidents, such as a dead tree falling into a herd of deer, kill at random. No selection occurs for age, size, sex, intelligence, and so forth. If, however, a predator kills the slowest member of a deer herd, that death was not random. Whatever trait made that deer slower is being selected against by the environment. Death in nature usually occurs by both paths. Whenever differential survival occurs, gradually the advantageous genes will increase in frequency in the gene pool.

This trend is best illustrated with an example. In a hypothetical population a dominant gene G is more advantageous to the species than the recessive gene g. Because of this the survival rate for the G_ genotype is 10% and that for gg is only 9%. If the starting gene frequency is 0.5 for each allele, then 25% of the offspring are GG, 50% are Gg, and 25% are gg. As shown in Table 13-1, the surviving offspring are composed of 25.64% GG, 51.28% Gg, and 23.07% gg. The resulting gene frequencies after only one generation are then 0.513 for G and 0.487 for g. If identical calculations are continued for 10 generations (Table 13-2) the gene frequencies become 0.606

Table 13-1 Calculation of Changing Gene Frequencies Resulting from Differential Mortality

Starting genotypes		0.500 for G	0.500 for g	
Multiply starting gene frequencies		0.500 G + 0.500 g 0.500 G + 0.500 g		
	0.250 GG	0.250 Gg 0.250 Gg		0.250 gg
Genotypes of offspring	0.250 GG	0.500 Gg		0.250 gg
Multiply offspring by 100 to put on basis of 100 offspring.	25.0 GG	50.0 Gg		25.0 gg
Apply assumed survival rates	25.0 GG 0.10	50.0 Gg 0.10		25.0 gg 0.9
Surviving offspring by genotype	2.50 GG	5.00 Gg		2.25 gg
Total of surviving offspring		2.50 + 5.00 + 2.25 = 9.75		
Calculate percentages of surviving offspring	$\frac{2.50 \text{ GG}}{9.75} = 25.64\%$	$\frac{5.00 \text{ Gg}}{9.75} = 51.28\%$		$\frac{2.25 \text{ gg}}{9.75} = 21.08\%$
Again multiply by 100 to put on basis of 100 offspring	25.64 GG	51.28 Gg		23.07 gg
Calculate the number of "surviving" genes		G = 25.64 + 25.64 + 51.28 = 102.56 g = 51.28 + 23.07 + 23.07 = 97.42		
Calculate percentage for surviving genes		$\frac{102.6}{200} = 51.3\%$	$\frac{97.42}{200} = 48.71\%$	
Gene frequency in new generation		0.513 for G 0.487 for g		

Table 13-2

Changing Gene Frequencies for Ten Generations in the Study Population of Table 13-1

Generation	Frequency for gene "G"	Frequency for gene "g"
Starting	0.500 G	0.500 g
2	0.513	0.487
3	0.525	0.474
4	0.538	0.457
5	0.552	0.447
6	0.564	0.436
7	0.575	0.425
8	0.586	0.414
9	0.596	0.404
10	0.606	0.394

for G and 0.394 for g. The actual difference in survival rates between organisms in a real population can be more or less than those given in this example, but the trend observed would be the same. That is, the frequency of the gene for the more advantageous trait increases if the individuals with that trait have a higher chance of survival (or reproduction).

Suggested Reading

SAVAGE, J. A. *Evolution*, Holt, Reinhart and Winston, New York, 1963.

Patterns in Evolution 14

The evolutionary process discussed in Chapter 13 focuses on the changes occurring within one species in a given environment. Some general overall evolutionary patterns also appear in a much larger area, consisting of many different kinds of environments and many different species. This larger environment coupled with the biological community of interacting species which it supports is called an ecosystem.

Competitive Exclusion

Within an ecosystem, each species assumes a particular role due to its interaction with the environment as well as its interactions with the other species. This is called the ecological niche for the species. It includes factors such as where the species lives, what it eats, what it does, and how it defends itself. For instance, a deer is a large, browsing, herbivore that is mobile, widely distributed throughout certain environments, and an important prey for large carnivores. If another animal, either a different form of deer or a different species with the same characteristics, entered the same environment, competitive exclusion would occur. That is, whichever form was better adapted to the total environment would tend to exclude the other from the ecosystem. The mechanism, again, for this overall pattern is natural selection. In this instance, however, one species may have a different survival rate than another species, rather than one form within a species having a different survival rate than another form.

Competitive exclusion is well illustrated by the following experimental results. Two different species of paramecia grow in the same aqueous medium. If their growth is studied separately, one species grows faster than the other. When equal numbers of the two species are put together in the same medium, the percentage of the faster growing species within the total population increases quickly. The slower growing species effectively disappears from the population within a short time.

Adaptive Radiation

Adaptive radiation, another overall pattern of evolution, occurs when a species enters an environment which has little or no pre-existing biological community or, at least, has some room for biological expansion. Since there is little or no competition for many ecological niches in this new environment, many modes of life are open to this newcomer. A species that is generally adapted to this new environment branches out to occupy many different ecological niches relatively quickly. Once these niches are established, natural selection insures that the forms within each ecological niche adapt specially to that niche.

The theory of adaptive radiation is based on fossil records and similarities found amongst present-day organisms. Fossil records, as far as they exist, indicate that some classes of organisms can be traced back to a common ancestor. The similarity of skeletons of the vertebrates indicates that the vertebrates developed from a common ancestor.

Adaptive radiation is indicated if the forelimbs of all classes of vertebrates, reptiles, fish, mammals, birds, and amphibians, are compared (Fig. 14-1). For each bone in the forelimb of any class of vertebrates, there is a corresponding bone in the forelimb of the other vertebrate classes. The shape and size of these bones differ, however, from class to class and from species to species. These differences reflect the adaptation of each vertebrate class to its mode of existence within its environment. For instance, the forelimb-span/body-length ratio is greater than 1 for birds, is much less than 1 for fish, and is close to 1 for terrestrial vertebrates. These differences are precisely those which allow the best adaptations of the basic vertebrate skeleton to the three different modes of locomotion required by the different ecological niches which these classes occupy.

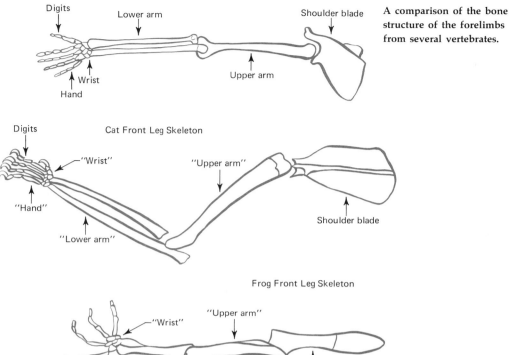

14-1 A comparison of the bone structure of the forelimbs from several vertebrates.

Divergent Evolution

Evolutionary trends in which two or more species develop from a common ancestral species are known as divergent evolution. Isolation-speciation and adaptive radiation are two such trends. An outline of the evolutionary relationships among organisms, which shows how species diverged into other species, is called a phylogenetic tree. In such a tree only the branch ends represent living species. The branch network merely indicates the genetic stock from which these species arose.

Phylogeny, the study of evolutionary relationships among organisms, is based on fossil records and the similarity of forms found among present-day organisms. All the organisms on earth today have, in general, evolved considerably from the primitive forms that radiated into many different ecological niches. As one species diverged into two or more other species, these new species were quite similar in form. With time, however, each of these new similar species would change and perhaps further speciate due to mutational and environmental factors. All of the species arising from a given species would again be quite similar, but that group might no longer be closely related to the forms that developed from a species originally similar to the one from which they developed. In general, then, species are more similar to one another the more recently they diverged from common stock. Since a phylogenetic tree shows the genetic history of all organisms, it also indicates which forms are more similar.

A highly diagramatic phylogenetic tree (Fig. 14-2) illustrates these divergent evolutionary trends. A species should, in general, be more similar to other species close to it on the phylogenetic tree than it is to those farther away. Because of this similarity in form due to evolutionary relationships, taxonomy, the system by which organisms are classified and identified, is based on the phylogenetic tree. As shown in Fig. 14-1, the taxonomic categories follow phylogenetic lines.

A close look at the individual species from divergent branches of the phylum chordata shows the increase in genetic similarities between two species the closer the evolutionary relationship is between them.

Man is a chordate, but so is the quite dissimilar lamprey. The lamprey is a parasite with a round suckerlike mouth that enables him to attach himself to fishes and feed on them (Fig. 14-3). Both man and the lamprey have the basic chordate characteristics: gill slits in the pharyngeal (throat) region, a dorsal hollow nerve chord, and a dorsal notochord. The lamprey has these characteristics all his life; man, however, only exhibits gill slits and a notochord in the developing embryo. The dorsal hollow nerve chord, however, is present in the adult man as well as in the embryo. Needless to say, the evolutionary trends leading to man and the lamprey diverged a long time ago.

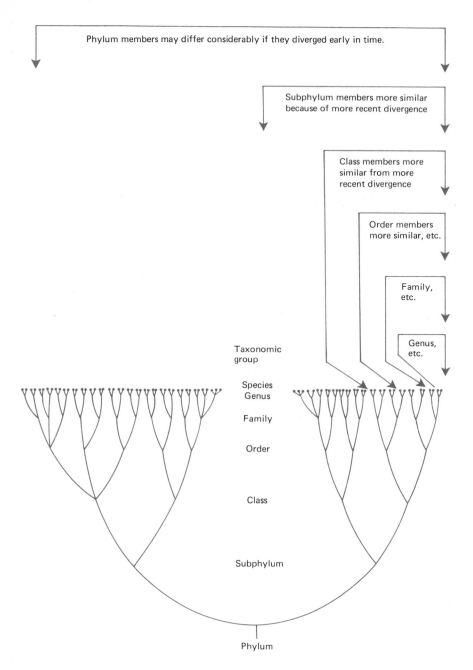

14-2 A diagramatic phylogenetic tree shows how taxonomic classifications follow the evolutionary relationships between organisms. In general, organisms are more similar to other organisms the closer the evolutionary relationship. That is, a species is more similar to other members of the same genus than to other members of the same family, order, class, subphylum, and phylum, in that order.

14-3 **The lamprey, a chordate like modern man.** Both man and lamprey have a simple supporting rod, a dorsal nerve cord, and pharyngeal gill slits at some stage of development. (*Courtesy of Wisconsin Conservation Department, Madison, Wisconsin.*)

14-4

A bird, a vertebrate form like modern man. All the similarities of the lamprey and man apply to the bird and man, in addition to the bony skeleton, vertebral column, and presence of two pairs of appendages. (*Courtesy of Colorado Game, Fish and Parks Division.*)

14-5

A rabbit, a mammal like man. All the similarities of the bird and man apply to the rabbit and man, in addition to the presence of mammary glands for nursing the young and hair on the body. (*Courtesy of Colorado Game, Fish and Parks Division.*)

14-6

A gorilla, a primate like man. All the similarities of the rabbit and man apply to the gorilla and man, in addition to the five-digited hands and feet with opposable digits, eyes that face forward, and flat nails instead of claws. (*Courtesy of Cheyenne Mountain Zoological Park, Colorado Springs, Colorado.*)

The bird (Fig. 14-4) is a chordate and also a vertebrate. The bird has more similarities to man than it does to the lamprey. It has a bony skeleton, a vertebral column, and two pairs of appendages. Although divergence between the stock that led to birds and that which led to man occurred later than that which led to the lamprey, men and birds differ fundamentally in many characteristics. Their similarities place them in the same subphylum, but their differences put them in separate classes.

A rabbit is a mammal as well as being a chordate and vertebrate. The rabbit shares with man all the similarities between man and the lamprey or bird, but it also shares additional characteristics, since men and rabbits are fellow mammals. These similarities are body hair and mammary glands for nursing their young. The differences are still innumerable, however (Fig. 14-5).

The gorilla is also a mammal, vertebrate, and chordate (Fig. 14-6). It shares with man the characteristics that man shares with the

Patterns in Evolution [191]

14-7

Pithecanthropus erectus, **a member of the family hominidae like man.** All the similarities of the gorilla and man apply to *Pithecanthropus* and man, in addition to upright stature and higher forehead that increases the brain capacity above that of most primates. (*Courtesy of American Natural History Museum, New York.*)

rabbit, the bird, and the lamprey. The gorilla and man, however, are also primates, so they both have eyes that face forward, hands and feet with five well-developed opposing digits, and digits with flat nails rather than claws.

Another primate, *Pithecanthropus erectus,* whose existence is indicated by fossils, is in the same family, hominidae, as man is. This primate shared the characteristics that man and the gorilla share, but also had the generally upright posture and higher forehead for greater brain capacity that man has (Fig. 14-7).

The Neanderthal man was a fellow member of the genus *Homo* with man. He had a greater brain capacity, a physical structure

14-8 Neanderthal man, a member of genus *Homo,* like modern man. All the similarities given for *Pithecanthropus* and modern man apply, in addition to a still greater brain capacity and tool making. (*Courtesy Field Museum of Natural History, Frederick Blaschke sculpture, Chicago, Illinois.*)

more like modern man, and used tools. *Homo Sapiens,* modern man, then, is more similar to this *Homo neanderthalensis* (Fig. 14-8) than to *Pithecanthropus erectus.*

Convergent Evolution

Certain structural forms allow an organism to adapt well to life in a certain type of environment. Because of this, some organisms, although distantly related genetically, are quite similar in form in one or several respects. This evolutionary trend is called convergent evolution. Generally, two different organisms which are similar in structure in the same or overlapping environments only resemble one another in a few characteristics. If these two species were similar in a large number of respects, they would both compete for the same ecological niche and one species would be eliminated through the mechanism of competitive exclusion.

Convergent evolution can be clearly seen in the similar overall body configuration of many swimming organisms. The similarities between the dogfish shark, the trout, and the porpoise are readily apparent (Fig. 14-9). All of these animals have a generally similar external form; yet, genetically, they are only distantly related within the animal kingdom. The dogfish shark is a cartilagenous fish. It has a skeleton made of cartilage, not bone. Cartilagenous fish are

14-9

Examples of convergent evolution. The dogfish shark, the trout, and the porpoise all have the grossly similar "streamlined" form, despite their rather distant evolutionary relationships. This is because all live in an aquatic or marine environment where swimming is a part of their way of life.

the result of a different evolutionary branch of the phylum chordata than the vertebrates. The trout and the porpoise, although both vertebrates, are also quite different. The porpoise is not a fish but a mammal.

Although these animals are obviously only distantly related genetically, their body forms are very similar. Because all these water animals must swim to find food and escape their enemies, they all have a body form which makes rapid movement through water possible. The streamlined form and the flat fins, which they all share, give them this capability. In the evolution of aquatic and marine forms from different genetic stocks, these characteristics were favored. Within diverging evolutionary trends, then, organisms with quite similar forms emerged because these forms best suited those individuals to their particular environments.

Similarity of forms amongst quite different organisms is quite common. For instance, the digging legs of the mole, a mammal, are quite similar to those of the mole cricket, an insect. The wings of a bird and a bat are also quite similar, in spite of the fact that the bat is a mammal and is covered with hair whereas the bird is covered with feathers. The hind legs of both an insect and man are so similar in form that the corresponding parts of each are given the same names. Genetically different organisms that are similar as a result of convergent evolution, are similar in only a few respects, as their ecological niches are different.

Convergent evolution leads to structures which are similar in function although the organisms which exhibit them have vastly different evolutionary origins. Structures such as these are called analogs. Homologs, on the other hand, are structures which are similar in form because the owner organisms have a common ancestry. Homologs are not necessarily similar in function, and they result from adaptive radiation. For example, the forelimbs of the vertebrates all possess similar skeletal features. Birds, however, use these forelimbs to fly, whereas horses use them to trot.

The Time Pattern of Evolution

The explanation of the origin of life and the subsequent evolution into the diversity of biological forms present on the earth today is theory. There is a great deal of evidence to indicate that the proposed mechanisms of evolution, natural selection, adaptive radiation, isolation-speciation, and so forth, can account for the diversity of organisms *given enough time*. The many random reactions and the many steps which must occur in the organization of simple chemicals into the complex organisms of today would require a great deal of time.

The most reliable method for determining the time periods involved in the evolution of the earth and living forms is radioactive dating. This process is based on the fact that certain atoms are structurally unstable and that, on the average, a certain percentage of those atoms decays to a more stable form within a constant amount of time. Such atoms are called radioactive. Uranium with atomic weight of 238, ^{238}U, was an atomic constituent of the early earth. It is radioactive and slowly decays to an isotope of lead, ^{206}Pb.

Ordinary chemical reactions do not involve the nuclei of the reacting atoms. The nuclei of radioactive atoms, however, are unstable. Three types of radiation can be emitted from the nucleus when a radioactive element decays. Alpha radiation (α rays) occurs when a radioactive nucleus gives off a particle consisting of two protons and two neutrons. Beta radiation (β rays) occurs when electrons are emitted from a radioactive nucleus after a neutron changes to a proton and an electron. An unstable nucleus can also emit gamma radiation (γ rays) which is similar to, but more powerful than, x-rays.

When ^{238}U decays to ^{206}Pb, all three types of radiation occur. Each decay occurs at random, but, on the average, as for any particular radioactive element, it always takes a constant period of time for a certain percentage of the atoms of a sample to decay. The time required for 1/2 of the atoms of a radioactive element to decay is called the half-life. The half-life for ^{238}U to decay to ^{206}Pb is 4.5 million years.

In radioactive dating, it is assumed that a deposit of a radioactive material was originally pure, and all breakdown is a result of radioactive decay. From the percentages of the radioactive material left within a sample, the age of the sample can be calculated. Using ^{238}U as an example, radioactive samples of the following composition would indicate the corresponding ages:

Sample compositions		Age, millions of yrs
1/2 ^{238}U	1/2 ^{206}Pb	4.5
1/4 ^{238}U	3/4 ^{206}Pb	9.0
1/8 ^{238}U	7/8 ^{206}Pb	13.5
1/16 ^{238}U	15/16 ^{206}Pb	18.0

Table 14-1

Approximate Timetable of Evolution

Origin of the earth	4,500 million years ago
Origin of "life"	2,000 million years ago
First "modern" cells	1,000 million years ago
Algae	350 million years ago
Reptiles dominant	200 million years ago
Mammals, birds and flowering plants	100 million years ago
Man	1 million years ago

Using radioactive dating from ^{238}U deposits and deposits of other radioactive elements, many geologic deposits have been dated. The age of the earth has been calculated by these and similar methods to be 4.6 billion years.

The timetable for evolution (Table 14-1) was proposed from radioactive dating, fossil records, and other methods of geologic dating. This timetable makes the theory of evolution seem plausible. It is conceivable that in 2 billion years life could have evolved basically along the proposed steps. The endless trial-and-error and selection processes necessary for biological evolution could have occurred over that time expanse.

Fossils represent the only real records of evolution. They have recorded the forms and sometimes the ages of former life. Unfortunately, fossil records are not complete or all inclusive. Only certain types of organisms could be fossilized, and certain environmental conditions were necessary for the process. The fossil record of the evolution of the horse is fairly complete, but that for many other organisms is poor to nonexistent. Where fossil records do exist, they substantiate evolutionary patterns and hence lend credence to the overall theory of evolution.

Suggested Readings

DODSON, E. O. *Evolution: Process and Product,* Reinhold, New York, 1960.
SAVAGE, J. A. *Evolution,* Holt, Rinehart and Winston, New York, 1963.

Part V

Energy Flow, Cybernetics, and Population Dynamics

Hawk eating mouse on fence post.
(*Courtesy of Colorado Game, Fish, and Parks Division.*)

Thermodynamics 15

All organisms require energy to live. The manner in which this energy is utilized within an organism to maintain its structure and organized functioning is perhaps the most wondrous aspect of living things. This becomes quite obvious against a background of the energy relationships encountered in mechanical systems.

Energy is a rather abstract concept. It can be simply, perhaps oversimply, described as the ability to do work. Energy seems a bit mysterious because it can only be described, perceived, or measured through its action on matter. It has both active and inactive forms. Kinetic energy is energy in action, whereas potential energy is energy that is inactive or being stored. There are many different forms of kinetic and potential energy. Chemical energy is the energy stored in the chemical bonds of a compound. Energy that travels in waves, such as heat, light, and x-rays, is called radiant energy. Electrical energy causes the flow of electrical charges. One form of energy can be interconverted to another form, and energy can be transferred from one object to another. Any interconversion or transfer of energy may be thought of as an "energy flow." Energy is most easily perceived during such an interconversion or flow of energy. An energy flow is manifest as a performance of work (movement of an object through space), or a flow of heat, or both. For instance, a match has potential energy in the form of chemical energy. This energy content becomes evident when the match is struck and the chemical energy is converted to heat and light as the match burns. The energy "flows" from the match to its surroundings, and the surroundings become noticeably warmer.

Thermodynamics is the study of energy and its interconversions. Studies of this kind are most conveniently undertaken in isolated systems. Neither matter nor energy passes into or out of an isolated system. Two basic laws, called by the austere titles of the first and second laws of thermodynamics, describe the way in which energy flows in an isolated system. A general knowledge of these two principles is helpful to the understanding of energy flow in biological systems like an organism or a biological community. Biological systems, however, are not "isolated" systems. They are "open" systems as both matter and energy pass into and out of them.

The First Law of Thermodynamics

In simple terms the first law of thermodynamics states that energy can be neither created or destroyed. This implies that during an energy flow there is neither a gain nor loss of total energy. This law holds for an energy flow through any isolated system, mechanical or biological.

Mechanical systems are notorious for their inefficiency. For instance, the amount of electrical energy produced by a generator powered by the burning of coal is not equal to the amount of chemical energy originally contained in the coal that was used for the conversion. This appears at first to be an exception to this first law which states categorically that energy is not lost. A careful study of the various steps of the energy flow from the chemical energy of the coal to the resultant electrical energy shows that no energy was "lost." At each energy interconversion, however, not all the energy was "harnessed." For instance, when the coal is burned to heat water and thereby produce steam, some of the chemical energy of the coal leaks out as light and some of the heat released is transferred to the surroundings rather than the water. Only the energy that heats the water to produce steam is "harnessed" to do the work of the generator system. The "unharnessed" energy is not destroyed in this step, it is merely unavailable to the practical task at hand.

In the next step in the production of electricity, the steam produced turns a generator. That is, the heat energy in the steam is converted to kinetic energy of the generator. Again, the entire energy content of the steam is not converted to kinetic energy; some of its heat is dissipated to the surroundings during the process, and some heat remains in the steam as it is exhausted from the generator.

As the kinetic energy of the generator is converted to electrical energy, some energy again gets distributed to the surroundings.

Some of the energy is converted to heat by the friction in the bearings of the generator or as the electricity produced travels through wires. The electrical energy produced can then be used in a number of ways, to produce light, heat, or motion. In each one of these interconversions some energy is converted to a different form than the light energy, heat energy, or kinetic energy being harnessed, respectively. This energy gets dissipated to the surroundings. In most cases the energy that escapes from the channeled system into the environment is in the form of heat (Fig. 15-1).

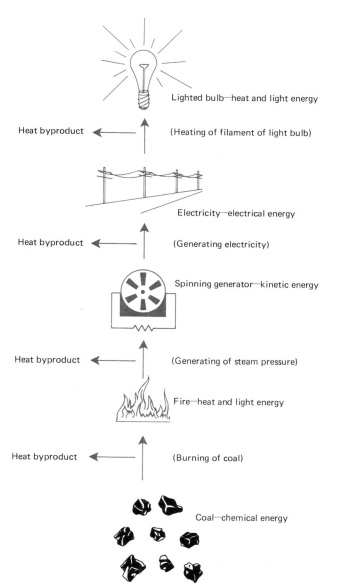

15-1

The flow of energy in the generation of electricity.

Thermodynamics [201]

This "inefficiency" of energy interconversions is typical of all cases of energy flow. It occurs within organisms and their communities as well. Organisms such as the human obtain energy in the form of chemical energy of their food. Within the human this food-fuel is "burned" slowly; the potential energy released is transferred to the energy of other chemical bonds, which maintain the chemical organization of the various compounds which compose body structure, and to kinetic energy of body movement. As in the generator, some of the energy appears as heat. Although some of this heat enhances the efficiency of many body functions, much of it is lost to the surroundings. The body temperature for many organisms, including man, is usually above that of their environment, and their bodies are constantly losing heat to the environment. Aside from the loss of heat to the surroundings, man also transfers to his environment the chemical energy of the chemicals which get passed out of his body and kinetic energy as he moves things in his surroundings. The mechanical efficiency of a human muscle, for example, is at best about 20–30%. The other 70–80% of the energy appears as heat rather than kinetic energy.

Energy flows in a similar manner through every organism in a biological community. The source of energy differs for each organism, as one organism supplies another with chemical energy in a "food chain." The ultimate source of energy for a biological community is light energy from the sun. Plants capture about 2% of the solar energy which hits them. This energy is converted into chemical energy during photosynthesis. Some of the chemical energy stored during photosynthesis is then slowly released within the plant and converted into heat energy and other forms of chemical energy which drive the reactions of the plant cells. However, all of the trapped light energy, no matter how many transfers it undergoes within organisms, is eventually transferred to the surroundings as heat.

The chemicals synthesized within a plant, which the plant itself does not use for energy to drive its own functioning, are available to serve as food for herbivores. This form of chemical energy is converted within the herbivore to other forms of chemical energy which drive the cellular functions of the herbivore, ultimately providing it with structure and movement. Again, much of this energy is surrendered to the environment as heat.

Carnivores feed on herbivores. The carnivores can use the chemical energy locked in the structural components of the herbivores to drive their own cellular functions by conversion of this chemical energy to other forms of chemical energy. Some energy escapes from the carnivore and heats up the environment during this conversion.

All these organisms—plants, herbivores, and carnivores—supply decay organisms (bacteria and fungi) with chemical energy. This energy, when released, maintains the decay organisms and also warms up the environment. When decay organisms die, their remains return to the soil. There is not much of the original, photosynthetically trapped chemical energy left in their remains, which eventually consist of only a few simple carbon compounds and inorganic compounds. These compounds get incorporated into energy-rich compounds in plants during further photosynthesis. The process of photosynthesis is, therefore, the basic input of energy into the biological community (Fig. 15-2).

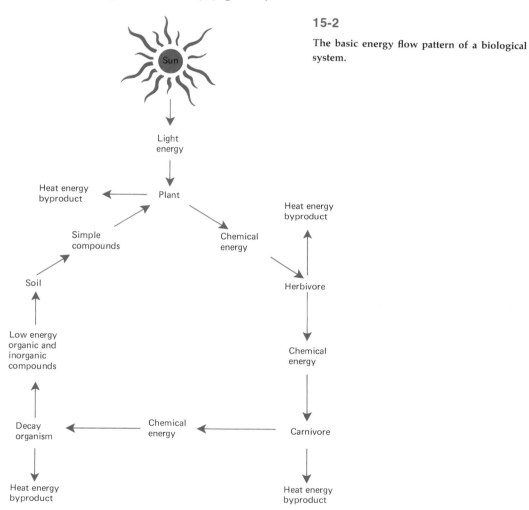

15-2

The basic energy flow pattern of a biological system.

Thermodynamics [203]

The Second Law of Thermodynamics

Any energy flow in a system involves a performance of work, or a flow of heat, or both. The second law of thermodynamics is the concept relating to energy flow which states in general the principle that, during any energy transfer in an isolated system, some energy is invariably converted to heat energy and the system loses some of its capacity to perform work. Again, this law holds for all isolated systems. The law was illustrated by each of the situations just analyzed. In each case heat was produced as energy flowed within the system, and the total energy was not all "harnessed" to produce work. The second law implies that some "loss" to "low grade" heat is, in fact, inevitable.

Energy flows only when there is a difference in energy levels within a system. Heat energy is measured by temperature, so heat energy can flow only when there is a difference in temperature within a system. If some of the energy of a system is converted to heat, a temperature difference is created within the system. This heat will "flow" throughout the system until the system is once again at a uniform temperature. Once heat flow occurs, the heat energy level difference in the system no longer exists. This energy can, therefore, no longer flow, and, hence, no longer perform useful work.

Entropy

The second law of thermodynamics, then, indicates that systems tend to "run down." As the energy of a system spontaneously gets converted to heat, it can do less useful work. In the attempts to describe thermodynamic systems mathematically, a specific mathematic quantity, called *entropy,* was defined. This quantity was formulated to simplify the mathematics of thermodynamics, but it also seems to have a physical meaning. Entropy is apparently a function of the state of a system, and it seems to "measure" the amount that a system has "run down."

There are different ways to try to conceptualize entropy in order to get an intuitive feel for what it measures. It can be considered as the degree of disorder in a system: the higher the disorder, the higher the entropy. It can also be considered as a measure of the distribution of energy within a system. The more diffuse the energy is in a system, the greater is the entropy. Using this quantity, the second law of thermodynamics can be restated as the tendency for entropy to *increase* spontaneously in an isolated system.

How the degree of disorder of a system can be a measure of the amount of energy converted to heat energy which can no longer do useful work arises from the nature of heat energy. Heat, most simply, is the *random* motion of the particles of matter (molecules, atoms, or whatever). In other forms of energy, the motion of particles is not completely random but "ordered" in some way. For instance, the major motion of all the particles of an object with kinetic energy is in the direction of motion. As another example, chemical energy holds atoms together so they move as a unit. In the sugar molecule, $C_6H_{12}O_6$, for instance, the 24 atoms have a particular arrangement in space relative to one another which is strictly maintained by the chemical bonds. These atoms are considered highly "ordered," since they have specific places within this structural array. Once such "ordered" motions are interfered with by friction or chemical reaction, the motions of some of the particles become random and, therefore, some of the order is lost. When a moving object collides with something that stops its motion, the particles of the object, which were all moving in one direction before impact, are all moving in different directions after impact. The entropy or degree of disorder has definitely increased during this energy interconversion. When a sugar molecule reacts with six molecules of O_2, six molecules of H_2O, six molecules of CO_2, and heat are formed. The atoms in these 12 new molecules are much less "ordered" with respect to one another than they were in the glucose molecule. The carbon atoms are no longer lined up in a row, but are free to move away from one another. The entropy of the atomic system has increased. On the macroscopic scale, then, an increase in entropy involves the conversion of other forms of energy, such as kinetic or chemical energy, to heat energy.

There are infinitely many more ways for the motion of particles to be disordered than ordered. Therefore, a disordered system is more probable than an ordered system. Alternatively, then, the second law of thermodynamics states that a system tends to shift from less to more probable states. To reverse this normal directionality of natural processes, energy must be supplied to the system.

Changes in which the entropy increases lead to more stable states. As the entropy increases, the system goes to a more disordered state and the energy is distributed over a larger area of the system. It is very unlikely that the system will become more ordered on its own or that this energy will be spontaneously reconcentrated. It is very improbable, therefore, that the change will be reversed. The final state, then, after a change which increases entropy is more stable than the initial state.

Although systems tend to "run down" during energy flow, the speed at which this occurs can vary. This speed depends on how the energy conversion is carried out. Once started, the reaction of $C_6H_{12}O_6$ with six O_2 to form the less ordered six CO_2 and six H_2O occurs spontaneously and gives off a lot of heat and light. In living organisms, however, this reaction occurs in several steps due to the intervention of enzymes. The reaction then occurs in more ordered steps; less heat energy is therefore lost to the surroundings, even though the total energy change is the same in both the spontaneous and the enzymatic conversions. In the enzymatic conversion some of the energy is captured as chemical energy as other chemical reactions, such as the formation of ATP, take place in the cell.

Open and Isolated Energy Systems

A system with no energy flow into or out of it is called an "isolated" system. The first law of thermodynamics indicates that within an isolated system the energy content remains constant. Although the amount of energy is constant within an isolated system, the second law implies that the amount of work available from this system gradually decreases.

A living organism is a highly organized system of very structured molecules. The second law implies that this system, if isolated, will tend to become less ordered and ultimately the chemical potential energy of these structured compounds will end up as heat energy. Such an isolated biological system then could not persist. This trend toward disorder can only be reversed by an input of energy. Since organisms obtain their energy from the break down of chemical compounds, a persisting organism cannot be a isolated system. Living organisms, since they require both an energy source and material input, are "open" systems. They continue to function only if they obtain a continuous supply of food energy.

Since each organism requires an energy supply, the biological community also requires an energy supply from the outside. An isolated biological community would very quickly run down. If plants which are the energy supply for herbivores die, the herbivores will also die. If the herbivores die, the carnivores die. If all these organisms die, the decay organisms will have no energy source and they, too, will die. In such an isolated biological community, then, the energy once present in the structured complex molecules which compose the organisms would eventually get evenly distributed throughout the environment, mostly in the form of heat. Only a random distribution of the simpler compounds formed in the decomposition of the more structured components of organisms would remain once an isolated biological community ran down.

If, however, energy is supplied to these living systems, the trend toward increasing entropy can be reversed. Any biological system, below the biosphere as a whole, whether it is an organism or a community, can only be maintained if it is an open system. The ultimate energy source for a biological community is the sun. Plants capture this light energy and synthesize the ordered, energy-rich compounds, the food-fuels, which maintain their structure and function. The food-fuels synthesized by plants ultimately maintain the structure and function of herbivores which depend on plants, of carnivores, which depend on herbivores, and of the decay organisms, which depend on all organisms.

16 Cybernetics and Systems Control

Any system which performs a function when supplied with energy must have a specific organization of its functioning parts. These functioning parts interact with one another to form a kind of "line of communication" between an energy source and the overall functioning of the system. The study of this communication between an energy source and the function it produces and of the various mechanisms by which this communication can be interrupted, speeded up, or slowed down is referred to as cybernetics. Cybernetic principles apply to any kind of functioning system, be it a mechanical system or a living system such as an organism or a biological community. Living systems perform many different functions in relative harmony. A knowledge of general cybernetic principles is helpful to an understanding of how the functions of living systems are controlled.

Systems Control in Mechanical Systems

There are three basic ways that mechanical systems are controlled: rigid control, negative feedback, and positive feedback. Biological systems frequently make use of negative or positive feedback. Since mechanical systems are generally less complex than biological systems, examples of these three types of control will be presented first in mechanical systems for simplicity. The more complex biological systems will then be analyzed in terms of these control mechanisms.

A rigid control system is one in which the rate at which the system functions is controlled directly by the rate at which energy is supplied from the outside or by a rate set by some independent mechanism. The rate at which one part of the rigid system functions determines the rate at which the next part in the line of communication functions. An automobile is a rigid control system of this type. The rate at which the energy (gasoline) is supplied is determined by the depression of the accelerator. The amount of gasoline in the cylinders regulates the speed at which the engine turns over. This regulates the speed at which the drive shaft turns. The rate at which the drive shaft revolves then determines the revolution rate of the rear axle and the speed at which the car moves forward. The movement of the car is its overall function, and it was specifically determined by the amount the accelerator was depressed. Another such system is a pendulum clock with an "escapement" mechanism which sets its rate.

The Rigid Control System

Any functioning system causes a change. In a feedback system the resultant change "feeds back" or influences the function of the system. A feedback system, then, must contain a sensor which reacts to the change caused by the system's function. By its interaction with this sensing mechanism the resultant change, or product of the system, controls the function of the system. In a positive feedback system the product influences the system so that it "does more of the same." The change or deviation from the starting conditions, therefore, increases in a positive feedback system. The negative feedback system, however, reacts to the system's output in quite the opposite way. A change in output directs the system to take such action as would eliminate the change. The negative feedback system maintains a fairly constant condition. These two systems are best described by examples.

Feedback Systems

Negative Feedback Systems Any feedback system must have a sensor that reacts to its own output (hence the term "feedback"). This sensor reacts to conditions in such a way that it increases the action of the system under one set of conditions and decreases it under others. An example of a negative feedback mechanical system is a thermostatically controlled heating system for a building. This system maintains a fairly constant temperature range for a building. The sensor of the thermostat control system is a bimetallic strip.

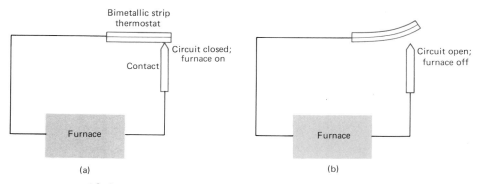

16-1 The negative feedback system as demonstrated by a thermostat. (a) Temperatures below the setting of the thermostat cause the bimetallic strip to straighten and close the circuit, turning on the furnace and raising the temperature. (b) As the temperature increases above the setting, the bimetallic strip curves; contact is broken, and the furnace is turned off. A relatively stable temperature is thus maintained in the house.

The inner layer of metal of this double-layered strip expands faster when heated than the outside layer of metal. As the temperature rises, the inside layer expands faster than the outside layer and the strip curves. As the temperature falls, this inner metal layer contracts faster than the outside layer and the curvature of the strip decreases. The bimetallic strip is arranged within the thermostat so that it closes the contacts of an electrical circuit when the curvature is low and breaks the circuit when the curvature is higher. When the circuit is closed, the furnace goes on; when the circuit is open, the furnace shuts off (Fig. 16-1). Thus, as the temperature rises above a certain point, the furnace goes off. The building being heated gradually loses heat to the surroundings and the temperature falls. At a certain point the circuit will be reformed as the bimetallic strip straightens, the circuit closes, and the furnace goes back on. In this system, as the furnace goes off and on, the building temperature oscillates between the upper and lower values of a small temperature range. In this way a relatively constant temperature is maintained within the building (Fig. 16-2). This is typical of the way in which a negative feedback control system can maintain a relatively steady state.

Positive Feedback A positive feedback system can be constructed if the bimetallic strip of the thermostat is turned over so that now the outer layer expands faster when heated than the inner

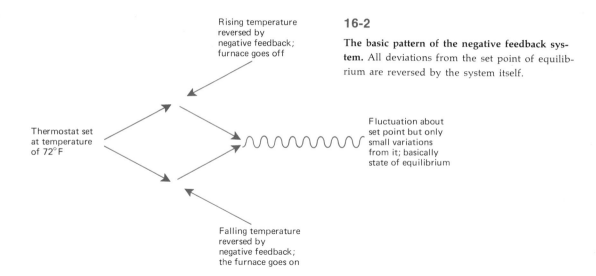

16-2

The basic pattern of the negative feedback system. All deviations from the set point of equilibrium are reversed by the system itself.

layer. The effect of a rise in temperature on this system is to close the contact and turn on the furnace and keep it on. A decrease in temperature, however, opens the circuit and turns off the furnace. The building cools even further as the furnace remains off. This system does not oscillate about a predetermined set point. Above a certain temperature it functions continuously, and below that point it does not function. The deviations from this point increase during the functioning or nonfunctioning of the system (Fig. 16-3). Since there is no balancing factor in this system, it is sometimes referred to as runaway feedback. Such a system goes automatically to maximum deviation from the switch point in one direction or the other.

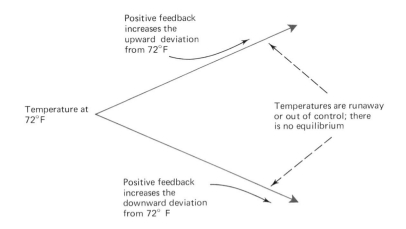

16-3

The basic pattern of the positive feedback system. All deviations from the set point of equilibrium are increased by the system itself.

Cybernetics and Systems Control [211]

Systems Control in Biological Systems

Organisms The enormous number of different natural processes which occur within biological systems, organisms, or communities would not operate harmoniously unless control mechanisms existed. The functional steady state that is observed in all organisms, either uni- or multicellular, is, in part, controlled by negative feedback mechanisms. The controlling mechanism responsible for the maintenance of a relatively constant human body temperature of 97–99° F is an example of a negative feedback control mechanism of a biological system.

The metabolic reactions which occur within the cell produce heat. This is the source of heat within the body. In general, heat increases the rate at which chemical reactions occur. If man had no mechanisms which controlled the amount of heat that his body loses to the surroundings, these reactions could, depending on the temperature of the surroundings, speed up or slow down. As metabolism occurs it produces heat. If this were not lost, it could increase the rate of metabolism which could produce more heat and even faster metabolism, and so on. Conversely, if a lot of heat were lost to the surroundings and metabolism slowed, less heat would be produced. This would slow metabolism even more and cool the cell still further. In short, the dependence of metabolism, a heat producing process, on temperature (heat content) would generate a rather unstable system with positive feedback.

There are, however, two basic ways which the body controls the amount of heat it loses to the surroundings. It can control the amount of blood that flows near the surface of the body by a factor of about 100. Heat is readily lost to the surroundings from the blood which flows to the skin. Hence, when the outside is cold, the amount of blood flowing to the skin is minimal to reduce this heat loss. When it is hot, however, more blood flows to the skin to increase the heat loss. The body also loses heat by producing sweat. Heat energy is necessary to change liquids into gases. Evaporation therefore cools a surface. The evaporation of sweat, then, cools the body.

The sensor which maintains a constant body temperature is in the brain. When blood that is colder than 97–99° F flows over this section of the brain, it triggers a decrease in heat loss by causing less blood to flow to the body surface and causing less sweat to be produced. If the blood that flows over the temperature-sensor area in the brain is warmer than 97–99° F, it triggers the mechanism for more heat loss; the body produces more sweat and more blood flows to the skin. In this way an increase in body temperature above 97–99° F feeds back to lower the body temperature by causing more heat loss. Body temperature below this range feeds back to decrease the heat loss and the body temperature rises.

Under severe conditions this negative feedback control mechanism is not effective. If the human is subjected to extreme cold, his body will continue to lose heat to the surroundings even after all its heat loss is minimized. When this happens and the body cools far enough below 97° F, the metabolism slows and less heat is produced within the body. The body then cools further. This process continues in a runaway feedback pattern and the body continues to cool. A body in this state will die unless it is carefully warmed from the outside.

Runaway positive feedback can occur in the opposite direction if the temperature control mechanism of the body gets upset and is set higher than the usual 97–99° F. This is the case during a "fever." Since the temperature control sensor is "set" higher than normal, the heat loss mechanisms are not operating at 97–99° F and the body temperature rises. Metabolism speeds up and the body gets still warmer. Generally the temperature control sensor is reset at a value which is not fatal to the body, and the body merely heats up to this value and maintains that temperature. If, however, the reset sensor value is very high, the body temperature will continue to rise. At a body temperature somewhere between 105 and 110° F the body can no longer cool itself at room temperature fast enough to offset the increased heat production due to the faster metabolism. The body temperature then will continue to rise in a state of runaway feedback.

Systems Control in Populations

The control of the size of a population of organisms can also be described as a negative feedback system. As seen from the reproductive potentials of all organisms in Chapter 12, the size of an animal population can run away with itself in a positive feedback system. For instance, the rabbit population resulting from one male and one female rabbit, assuming an average litter of four rabbits and no deaths, grows to 162 rabbits in only five generations (Table 16-1). The population triples in every generation. Because each output (rabbit) provides input by breeding more rabbits, this system is a fine example of runaway feedback.

In the biological community, however, one population of orga-

Table 16-1 Theoretical Population Increase Table for Rabbits

Assume no mortality, a litter size of four, and 50:50 sex ratio, for example, 2 + 2 = 4.

Generation	Age groups					Total population
	Young	1	2	3	4	
1	1 + 1 = 2					2
2	2 + 2 = 4	1 + 1 = 2				6
3	6 + 6 = 12	2 + 2 = 4	1 + 1 = 2			18
4	18 + 18 = 36	6 + 6 = 12	2 + 2 = 4	1 + 1 = 2		54
5	54 + 54 = 108	18 + 18 = 36	6 + 6 = 12	2 + 2 = 4	1 + 1 = 2	162

nisms does not exist in isolation. The control that the environment imposes on the size of a population of organisms is, in part, one of negative feedback. Using the same example, rabbits fall prey to their predators. As the number of rabbits increases, the number of predators also increases since they have more food (rabbits) available to them. Once the number of predators increases, however, the number of rabbits starts to decrease as more rabbits fall prey to these predators. As this trend continues, the number of predators then decreases since there are then fewer rabbits on which to feed. This predator decline makes the rabbits less threatened. Since there are fewer rabbits, however, each rabbit has more food and cover. The rabbits then start to thrive and increase in numbers again. In this zig-zag fashion, the rabbit population of an area oscillates around a relatively stable long-term average (Fig. 16-4).

16-4

Negative feedback system working in an animal population. This is the general pattern of certain population controls, although oversimplified here.

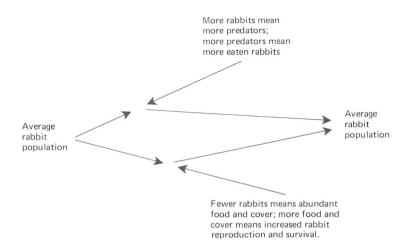

More rabbits mean more predators; more predators mean more eaten rabbits

Average rabbit population

Average rabbit population

Fewer rabbits means abundant food and cover; more food and cover means increased rabbit reproduction and survival.

Energy Flow, Cybernetics, and Population Dynamics

Animal Population Dynamics 17

The size and character of an animal population is determined by the interactions of that species with its environment. By these interactions an organism obtains from the environment the necessary raw materials and energy to grow and reproduce. An environment is the sum total of things in an organism's surroundings, living or nonliving, which influence the organism or its population. The energy requirements and the reproductive potential of a species are the two basic factors which shape its population within a particular environment. These control mechanisms can be easily demonstrated by analyzing the forces which shape the population of a particular animal such as the deer (Fig. 17-1).

Reproductive Potential

In an environment with an unlimited energy source and supply of necessary raw materials, an organism reproduces in a runaway feedback pattern, as previously noted. That is, with time, the animal population grows at a faster and faster rate. The theoretical deer population which could arise in 10 generations from one male and one female, if they (and any pair of progeny) produced a male and a female fawn every year for 3 years, is presented in Table 17-1. In the first 5 years the deer population would increase from 2 to 28 deer (1:14). In the next 5 years the population rises from 28 to 596 (about 1:21). Such increasing reproductive rates cannot continue indefinitely in a real environment without the number of organisms ultimately exceeding the carrying capacity of the environment.

17-1

The mule deer of the West. This is the most common deer of the western mountains. (*Courtesy of U.S. Forest Service.*)

17-2

Mule deer range in the West. The mountains are the summer range, while the deer commonly move down into the lower hills and valleys for the winter. Seasonal migrations are typical of this type of country. (*Courtesy of Denver Public Library, Western Collection. Photograph by Louis C. McClure.*)

Generation	Age groups				Total population
	Fawns	1	2	3	
1	1 + 1				2
2	1 + 1	1 + 1			8
3	2 + 2	1 + 1	1 + 1		8
4	4 + 4	2 + 2	1 + 1	1 + 1	16
5	7 + 7	4 + 4	2 + 2	1 + 1	28
6	13 + 13	7 + 7	4 + 4	2 + 2	52
7	24 + 24	13 + 13	7 + 7	4 + 4	96
8	44 + 44	24 + 24	13 + 13	7 + 7	176
9	81 + 81	44 + 44	24 + 24	13 + 13	324
10	149 + 149	81 + 81	44 + 44	24 + 24	596

Table 17-1

Theoretical Population Increase—Table for Deer

Assume twin fawns; reproduction at 1, 2, and 3 years old and death after 3; sex ratio of 50:50.

Energy Requirements

All organisms require energy to maintain life, growth, and reproduction. In any "successful" biological community, there exists a dynamic system of interdependencies between the organisms as well as with their nonliving surroundings such that each organism obtains the energy and raw materials it needs to maintain itself. This system of interdependencies is sometimes referred to as the "food chain" since "food" (energy-rich organic compounds) for most organisms supplies both the necessary raw materials to build cell structure and the energy necessary for this synthesis.

The food chain can be described as having three, sometimes four, trophic levels. The food producers, the green plants, occupy the first trophic level. The primary consumers, the herbivores or plant eaters, are the second trophic level. The carnivores who live on herbivores are the secondary consumers, and they compose the third level. The fourth level are carnivores who eat the secondary consumers.

The food chain can be illustrated by an analysis of the organisms on which the deer depends and the organisms which depend upon the deer for their requirements. The deer, a herbivore, lives on plants in its surroundings (Fig. 17-2). The plants synthesize the compounds which the deer requires during photosynthesis; they are able to utilize energy from sunlight to build up these organic chemicals from the simple inorganic components of the environment including CO_2, H_2O, and other simple compounds. A deer cannot perform this chemical operation and relies on the plant's ability to do so. The deer, however, utilizes only a small amount of the energy that the plant "fixes" from the sun into the organic food-fuels. The plant itself uses most of the foods that it synthesizes to maintain its own vital functions. Because of their photosynthetic ability, plants can obtain the energy and raw materials for their own synthesis of cellular constituents, active transport, cell division, growth, and other more specialized functions.

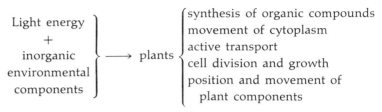

The chemicals composing the plant at the time it is eaten by a deer contain the energy source and raw materials needed for the deer's cell processes. From these food-fuels of the plant, the deer synthesizes cellular components and structures which are different than those of the plant. The deer is composed primarily of protein, the plant of cellulose. Most of this food, however, is "burned" within the deer to maintain its vital processes. The fact that a deer can keep eating without changing its own weight indicates that the foods are being broken down in energy-releasing reactions to smaller compounds, such as CO_2 and H_2O, which the body can eliminate. Many of the processes, such as synthesis, cytoplasmic motion, active transport, cell division and growth which also occurred in the plant, are typical of most cells. Some of the energy-requiring processes of the deer, however, are different from those of the plant. For instance, the deer moves around seeking food which it then digests; plants live a much more passive existence.

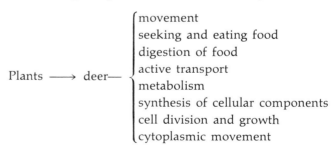

The deer, however, provides food for several types of animals, parasites, disease agents, and predators. The predator kills the deer and feeds on its body components. The parasites and disease-causing agents are usually much smaller organisms which live within the deer and feed on some of the constituents of the deer's cells without killing the deer. The liver fluke is a typical parasite; it uses compounds from within the deer's liver as both the raw materials and energy source for maintaining its own functions.

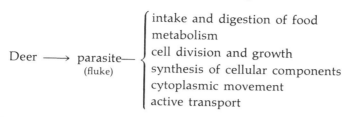

Deer ⟶ parasite— (fluke) { intake and digestion of food / metabolism / cell division and growth / synthesis of cellular components / cytoplasmic movement / active transport

Bacteria are typical of disease-causing agents. These organisms feed in and around the host deer cells and may cause an abnormal body function (disease) in the host. The processes which bacteria support are similar to those of most organisms.

Deer ⟶ disease agent— (bacteria) { movement / metabolism / synthesis of cellular components / cell division

The mountain lion (Fig. 17-3) is a major predator which threatens the deer. It feeds on the chemical components present in the deer at the time of the kill. The energy stored in these components, of course, represents only a small portion of the energy sources which the deer has utilized during its lifetime. The predator requires this energy to maintain its mobile food search, digestion, and other typical cellular activities. In spite of the difference in the mode of obtaining an energy source between this predator and its herbivore prey, both organisms utilize energy sources in much the same way.

17-3

The mountain lion. This animal is a very important predator of the deer where the two species occur together. (*Courtesy of Colorado Game, Fish, and Parks Division.*)

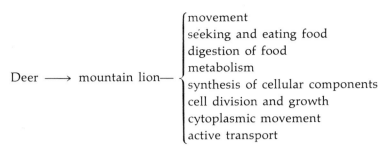

There are some organisms, bacteria, fungi, and others, which use the dead bodies of other organisms for a food source. These organisms are called decay organisms. There are so many different types of decay organisms that almost all organisms fall prey to one type or another decay organism after they die. This includes plants, herbivores, parasites, disease agents, and carnivores. The decay organisms use the chemical components of these dead organisms to maintain their structure and function in the same basic ways that other organisms do.

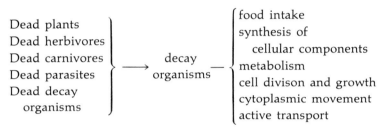

There are many modes of life, therefore, that organisms have adopted in order to assure themselves of a food source. By such adaptation, organisms have specialized in the utilization of particular energy sources. Although organisms have adapted to utilization of different energy sources, they all require an energy source. This need for energy for such similar purposes within the cell is an underlying unifying link amongst all living things, no matter how diverse.

Energy Flow and Population Size

Each species has its own characteristic energy requirements and reproductive potential. The character of each population within a community will depend on how successful each species of the community is in obtaining energy to maintain its own reproduction in that ecosystem. The overall energy flow through a biological community, then, shapes the population of each species within that community.

During the energy flow through the food chain, a great deal of energy is lost to the surroundings as heat at each trophic level. In general, only about 10% of the energy used by each level is transferred to the next level. This means that, on the average, the food value of any animal is approximately 1/10 the available food value of all the food that animal ingested in its lifetime. During any period of time, then, the energy taken in by the primary consumers is 1/10 that taken in by the producers. Similarly, the energy taken in by the secondary and tertiary consumers is 1/100 and 1/1000, respectively, that taken in by the plant producers. A diagram comparing the energy used by each trophic level necessarily forms a pyramid (Fig. 17-4).

Although the size of a population depends on the energy available to the species, the population sizes of the various trophic levels do not necessarily mimic this energy pyramid. The populations within a community are compared in terms of the number of organisms or the biomass (dry weight) of each population at a particular time. These comparisons are only concerned with the quantity of organic material present at any one time and do not take into account the total amount of material produced or the rate at which it is being produced.

Therefore, the relative numbers of organisms in each trophic level of a community vary depending on the size of the organisms. If the primary producers are small, a large quantity of them are necessary to support each consumer that feeds on them (Fig. 17-5a). If the producers are large, however, like trees, each producer can support a number of consumers (Fig. 17-5b).

17-4 A schematic diagram of the energy used by the different trophic levels in a specific ecosystem over a certain period of time. The length of each bar represents the energy of a trophic level. Code: P = producers; H = herbivores; C-1 = carnivores, secondary consumers; C-2 = carnivores, tertiary consumers.

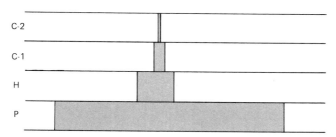

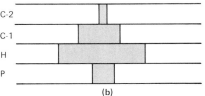

17-5 A schematic diagram representing two possible food chain relationships. Each bar represents the number of organisms within each trophic level of a particular ecosystem at a given time. Code: P = producers; H = herbivores; C-1 = carnivores, secondary consumers; C-2 = carnivores, tertiary consumers. (a) If producer organisms are small and/or unproductive, large populations of them are necessary to support consumer organisms. (b) If producer organisms are large and/or productive, each one can support several consumers and, at least, a partially inverted pyramid is possible.

The relationships between the biomasses of the trophic levels within a community also vary depending on the nature of the organisms within that community. If a producer can reproduce itself much more rapidly than its consumers can, the biomass of the producer at any given time may be less than the biomass which feeds on it (Fig. 17-6a). It is more often the case, however, that the biomasses form an upright pyramid (Fig. 17-6b).

The amount of energy that an organism in one trophic level fixes within its structure is only a small amount of the total chemical energy it takes in from organisms in lower trophic levels. As this chemical energy flow from one level to the next decreases, simple compounds are being regenerated which are necessary for photo-

17-6 A schematic diagram representing two possible biomass relationships in a food chain. The length of each bar represents the biomass (dry weight) of all the organisms within a trophic level of a particular ecosystem at a given time. Code: P = producers; H = herbivores; C = carnivores. (a) If the producers can reproduce themselves faster than the consumers can, the producer biomass may be smaller than that of the consumer. (b) In general the biomass of each trophic level is less than the biomass of the next higher level. This trend mimics the energy relationships.

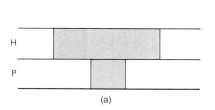
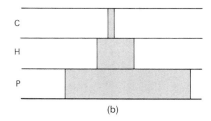

Energy Flow, Cybernetics, and Population Dynamics

synthesis and other organic syntheses. The continuity of a biological community relies on this regeneration of these simple compounds. The flow of energy, then, gives a biological community its organization, form, and continuity.

Systems Control Factors in a Biological Community

The ultimate force controlling the various animal populations is frequently the availability of the energy sources which each population can use. A particular population increases when more of its energy source is available per organism. However, each population serves as an energy source for other dependent populations. An increase in one population means an increased energy source for the dependent populations. This allows an increase in the size of the dependent population. As a dependent population increases, it uses more energy. The fulfillment of this need causes a decrease in the population serving as energy source. By virtue of this food chain, therefore, a population increase can trip mechanisms which reduce the likelihood of further increase. The variation in population size resulting from this negative feedback mechanism differs from species to species. The variation depends upon the particular interactions of each species with its total environment. This negative feedback mechanism and the type of variation it causes in the populations of various organisms can best be illustrated by analyzing the different factors which shape their populations.

The deer, for instance, relies on plant life for its energy source. The plant life within a region, however, depends on the physical conditions—temperature, moisture, and available nutrients—as well as the biological factors of disease, competition for light, and herbivores. In a particular region the plant species present must be those already adapted to the physical conditions so these will not be discussed. In general, plant communities have also adjusted to disease factors within the environment so diseases do not often cause fluctuations in plant populations. However, if a new disease is introduced into an environment, gross fluctuations in plant populations can result. For instance, the chestnut blight virtually wiped out the American chestnut, formerly an important part of the American hardwood forest. Similarly, the American elm is currently being severely threatened by the Dutch elm disease.

Competition between plant species in a particular environment is a major force in shaping plant populations in a particular area.

Plant communities are composed of the species that have competitively established themselves under the prevailing environmental conditions. The grassland is successful where it is and the forest is successful where it is. This tends to be a relatively stable situation.

Herbivores, however, interact with plant communities in a runaway feedback pattern until the plant populations and herbivore populations are almost totally threatened. There is, in effect, a long time lag before negative feedback comes into play. The plants provide a rich energy source for herbivores such as deer. The deer thrive and produce more deer. The offspring thrive and produce many more deer. This accelerating growth pattern will continue if the deer population is not checked by other means until the plant life of the area is decimated and thus introduces, at last, some negative feedback. The plant life can only be revived when the number of herbivores decreases. This may ultimately be caused by starvation of the animals.

This pattern was unfortunately realized in the Kaibab National Forest in the early 1900s. In an effort to increase the small deer herd, the area was made a game refuge and predatory animals were trapped. The deer population increased from an estimated 4000 in 1905 to 100,000 in 1924. Because of this bulging population, the range was ruined between 1924 and 1930 and approximately 80,000 deer died of starvation.

Deer populations, like those of other herbivores, are controlled by predation, parasitism, disease, and man's hunting as well as by plant supply. Parasites and diseases are generally not critical controlling factors in deer populations, however. These invaders are not usually a threat to healthy deer. If the deer are already weakened by other factors, such as starvation or exposure, disease agents or parasites may then cause their death.

The two prime factors controlling deer populations are food supply and predators. Neither factor works alone, but they will be described separately. Both factors exert their control in basically a negative feedback pattern. As already described, the control exerted by the food supply is fairly insensitive and permits temporarily large fluctuations in herbivore populations. In a rich food supply deer thrive and reproduce rapidly. Ultimately they could overproduce so that the food supply can no longer maintain the population. In an extreme case, many deer may starve if the production is too great, as demonstrated in the Kaibab range. As the food supply gets short, the rate of increase in deer population slows. The deer population decreases until it reaches a level that can be supported by the food supply. As the deer population decreases, however, the food supply has a better chance to grow because the deer population that was decimating it has shrunk. There will ultimately be a point where the decreased deer population has abundant food and cover and it will start to increase once again.

Predators exert the same type of pressure on the deer population, except the effects are not as severe or as immediate. As the deer population increases, the number of predators which live on deer also increases. Predators have a better chance at surviving and reproducing when food is abundant. This increase is not rapid enough to show up as sharp fluctuations in the predator populations, but slow population adjustments do occur. As the predator population grows, however, the deer population decreases because these animals are used up faster by the predators. Once the deer population has decreased to a certain extent, deer become harder and harder to find, and the predator population decreases. The deer then have abundant food and cover and their numbers increase.

The changes in deer population due to man's hunting may be controlled quite closely. Under ideal conditions, the size of the deer population is surveyed each year and a proportional hunting quota can be set so that hunters can help to maintain a certain stable population of deer. In this day and age the control of animal populations by hunting can be critical because in many places the number of natural predators has been greatly reduced. Without heavy predation or the intervention of man in a controlled manner, food supply will be the ultimate controlling mechanism of animal populations. Enlightened hunting, then, can fill the role of the natural predator and can prevent starvation in deer populations and the concomitant ruination of the grazing ranges. Sport hunting, in short, need not be the barbaric epitome of anticonservationism as its most extreme detractors sometimes suggest.

18 Human Population Dynamics

The factors which shape the populations of other animals also work to shape the populations of humans. However, superimposed on these forces are all those that result from the special social organization of humans and our decision-making ability. Man has been able to intercede in the various natural mechanisms of population control so that, to an unusual extent, he controls his own environment rather than being at its mercy. As man, however, is still subject to the pressures of reproductive potential and energy requirements, this control may prove incomplete, particularly if the population increases faster than man can shape his environment to support this population.

Pressures of Reproductive Potential

Man, like all other organisms, enjoys a great ability to reproduce himself. The reproductive rate of humans can be illustrated in a population increase table. (Table 18-1) For this table it is assumed that each couple has four children, two boys and two girls. Furthermore, the average reproductive age is assumed to be 25 and that for death is assumed to be 75. The population would theoretically change from 2 to 60 in five generations and from 2 to 7680 in 12 generations, or 300 years. The population would thus double every 25 years, a rate of increase currently met in many countries. The reproductive rate used in this calculation is not by any means a maximum rate yet this rate of 1 to 3500 in 300 years cannot be sustained for very long in any biological community. This increase must be limited somehow before the resultant human population comes into sharp conflict with its surroundings.

Table 18-1 Population Increase Table for Humans

Assume starting population of two, families four children, mortality after 75, 50:50 sex ratio, and brother-sister marriage.

Generation	Age group				Total population
	Child	25	50	75	
1	1 + 1				2
2	2 + 2	1 + 1			6
3	4 + 4	2 + 2	1 + 1		14
4	8 + 8	4 + 4	2 + 2	1 + 1	30
5	16 + 16	8 + 8	4 + 4	2 + 2	60
6	32 + 32	16 + 16	8 + 8	4 + 4	120
7	64 + 64	32 + 32	16 + 16	8 + 8	240
8	128 + 128	64 + 64	32 + 32	16 + 16	480
9	256 + 256	128 + 128	64 + 64	32 + 32	960
10	512 + 512	256 + 256	128 + 128	64 + 64	1920
11	1024 + 1024	512 + 512	256 + 256	128 + 128	3840
12	2048 + 2048	1024 + 1024	512 + 512	256 + 256	7680

Energy Requirements of Humans

Man requires energy for both biological and social needs. Because of his adaptability and ingenuity, man can control some of the ways he interacts with his environment in search of this needed energy.

Man has many of the same energy requirements as other animals have to maintain his biological functions. Since he cannot carry out photosynthesis, he requires a chemical energy source to provide him with necessary energy for movement, digestion, metabolism, synthesis of body components, cell division, and growth. He obtains these energy-bearing compounds from plants and other animals, and is therefore an omnivore.

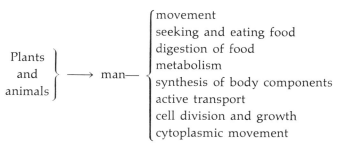

In addition, the maintenance of human society requires energy in much the same way as the maintenance of the human organism.

Industrial complexes use energy in the fabrication of products which society uses. Vast quantities of energy are used in transporting members of society and their products through the social complex. Humans are constantly reshaping the landscape, tilling the soil, cutting down trees, building roads, and so forth. Energy is required for all these projects. Energy is used in the construction of buildings which are then constantly supplied with more energy so that certain functions can be performed within these buildings. The energy sources used to maintain these and many other processes are widely varied. Man has ingeniously been able to harness energy from various sources such as water, wood, oil, coal, plants, animals, and atoms.

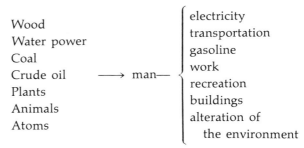

Through the social history of man, there have been different ways in which man interacted with his environment in order to obtain his energy requirements. A look at five "theoretical stages" of man's social development, the scrounge, the hunter, the farmer-hunter, the industrialist, and the atom splitter, shows how these different stages allowed for different sizes of human populations to be supported by the environment.

The "scrounge" is the primitive man who used little in the way of tools. His life must have been precarious and difficult amongst the other wild animals because he probably had nothing better than a stone ax with which to protect himself and gather food. He probably lived mostly off nuts, berries, fish, bulbs, shellfish, or anything else he could scrounge with just an ax, stick, stone, or fingernails. Since he was relatively helpless against many large and powerful animals, and had rather primitive methods of finding energy sources, the "scrounge" did not flourish.

The "hunter" developed tools such as the spear and the bow and arrow. Both these implements are effective in killing fairly large animals. By their use the hunter could not only defend himself from larger animals but could also kill them for his own energy source. His existence was therefore not so precarious, and there were probably more hunters than there were "scrounges." Carnivores, such as this "hunter," however, exist in relatively small numbers since they are dependent upon an energy source which is not in as abundant supply as plants.

The "farmer-hunter" grew his own plants. With this development he was able to obtain a food supply very efficiently. Herbivore populations grow to larger sizes than carnivore populations, since the energy fixed by the plants is used directly by the herbivore and only indirectly by the carnivore. This farmer was primarily a herbivore. However, man has never completely converted from hunter to farmer; he still requires meat intake for the amino acids needed by his body that plants do not produce. By taking up farming, however, man could make the environment support larger numbers of people. This became especially true as the farmer improved his agricultural techniques.

With the industrial revolution man could hunt, farm, and build more efficiently. He devised ways to produce tools, machines, electricity, and many other products by using the vast coal and oil reserves of the earth. With this advance man had once again increased the ability of the surroundings to support human population. His farms became more efficient and productive with the development of special farm equipment, so more people could be fed. Formerly uninhabitable areas were reshaped by industrial techniques so that the area open to human populations expanded as well.

The atomic revolution may produce a society where almost unlimited sources of energy are available. It is too soon to tell what this might mean in terms of changing the environment's ability to support human populations.

The ability of the environment to support animal populations is commonly called the carrying capacity of that environment. As man changed his mode of living and developed new ways to tap the energy sources of the earth, he changed the carrying capacity for humans on earth. The size of the human population of earth then has followed a series of jumps which reflect these changes in the living modes of humans (Fig. 18-1).

Environmental Changes Caused by Energy Use

The human population taps just about every energy source available except the sunlight itself. By virtue of these energy drains, however, man has altered his environment to a much greater degree than has any other species.

18-1

The theoretical steplike changes in carrying capacity of the earth for human populations as the mode of living for man changed.

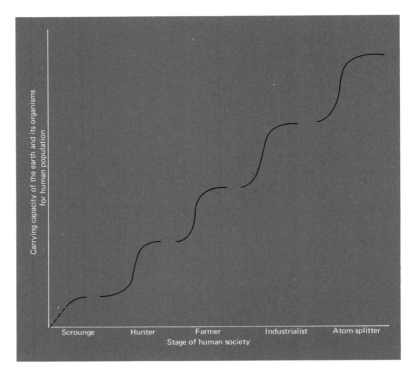

For instance, man has domesticated both plants and animals for use as food energy sources. Under natural conditions the environment is fairly stable since all the complex interactions between the environmental components are fairly stable. Man, however, does not utilize the naturally occurring plant populations. Instead, he breaks the soil and plants his own crops. Man's alteration of the plant community sometimes causes severe damage to the planting site. His domestication of animals can also cause the same destruction. Wild animals move about freely and rarely destroy a single plant community. Once animals are fenced in, however, they can easily overpopulate their grazing area and destroy the pasture. Despite the complete dependence of human populations on the plant community for food, the effect of human activities has sometimes been disastrous to the natural plant communities.

The wild predator populations have also been altered by man in his attempt to obtain energy. Predators compete with man for other herbivores, wild or domesticated. Man has killed off wild predators to make herbivores more available to man for his own use. Predators such as the timber wolf and mountain lion which are now quite rare in North America were killed off for this reason and not primarily because they were a direct threat to human life.

The taps which man puts on the other energy sources puts great stress on the environment. Coal, oil, and wood are all products resulting from fixed light energy from past photosynthetic biological communities. They are all being used in great quantities. Wood is being used extensively in construction and in paper manufacture. It is being cut down faster than it can be grown. Coal is used at a rapid rate to heat buildings and generate steam power. Great quantities of oil are being used to produce gasoline and other products. All these commodities produced by past generations are being used at rates that are faster than their current production within the natural community.

Most organisms only use energy that other organisms are currently producing. Man, however, not only lives on energy that is currently being produced by other organisms but also uses up the earth's reserves for his social needs. The large energy use of man can only continue if another energy source such as atomic energy is readily available once the coal and oil energy reserves are depleted.

Not only is man using up the energy reserves of the environment, but he is also using up the raw materials as well. The chemical components of other organisms are decomposed to smaller compounds that return to the environment when these organisms die. These compounds eventually get incorporated into new organisms. The components of humans, however, do not get recycled in this way. Their bodies are embalmed and buried in cement vaults. Decay organisms can no longer break down these remains to the simple chemical compounds which plants can resynthesize into energy foods. The glass, brick, iron, and concrete structures which man builds are also impervious to such decay. Of more importance is the fact that in his careless use of the land, much of the raw materials formerly present in the soil have effectively been lost. This has occurred either by depletion due to poor farming practices or by erosion. Much of the rich topsoil has been washed away into the ocean due to man's alleged "control" of the environment.

Man makes many more demands on the environment for his social and biological needs, therefore, than most animals who only seek to fulfill their biological needs (Fig. 18-2). Man's use of the environment can be very destructive to the earth and the biosphere it supports. The responsibility for maintaining the environment for his own use as well as for other organisms rests solely on himself.

18-2

Energy flow into human populations. Arrows show the direction of energy flow.

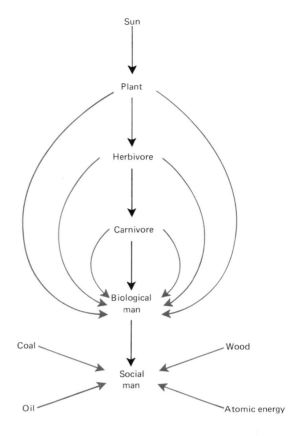

Systems Control Factors in Human Populations

The reproductive potential and inflated energy requirement of the human population may damage the entire biological community as well as the human community. Man seeks to control any of the natural forces which tend to control animal populations such as food supply, parasites and disease, and predation. Removal of the immediate environmental control of his population implies that he will have to substitute some other control in order to prevent a runaway feedback pattern of population growth which would ultimately result in a break down of the entire biosphere. Other controls have not yet proven effective, and the world population is doubling about every 35 years. The following discussion about modes of control which man, himself, could impose on his population is focused entirely on the biological aspects of the problem. It does not touch on ethical or moral considerations.

Of the natural control factors which counteract the reproductive

[232] Energy Flow, Cybernetics, and Population Dynamics

potential in other populations, parasites, disease, and food supply are the most important to the human population. Human population control due to weather and predation has not been significant for some time. Parasites have been, and still are, an important control factor. Malaria, the number one disease in the world, is caused by a tiny (protozoan) parasite which lives in human red blood cells. This disease has caused millions of deaths and illnesses. Another protozoan parasite causes African sleeping sickness which is often fatal. Worms and many other parasites disable and kill people the world over, every day. The effect of parasites as a population controllant, however, is lessening all the time. Medical research has found ways to control these parasites, and these medical techniques have spread to many areas. If the ultimate goal of controlling all these parasites through research, sanitation, and medicine is ever realized, parasites will no longer exert even the slight control which they have at present.

Diseases have also been effective population controllants. The bubonic or black plague killed 1/5 of the people of continental Europe during one period of time. Yellow fever was once the scourge of the tropics; typhoid fever and cholera were often fatal. Other diseases such as small pox, whooping cough, poliomyelitis, scarlet fever, diphtheria, tetanus, dysentery, and pneumonia all took many lives, especially among children. Some of these diseases are still important in some areas of the world. Again, medical research and the spread of medical know-how is reducing the effect of most of these diseases. Immunizations now protect people from small pox, tetanus, whooping cough, red measles, diphtheria, yellow fever, plague, typhoid fever, polio, and other diseases. Antibiotics are effective in the treatment of still other diseases such as pneumonia, scarlet fever, infectious dysentery, and, in some cases, syphilis. These diseases will lose their population controlling effect if the goals of medical research are realized and their results universally implemented.

If there are no other controls imposed on the human population, the food supply will be the ultimate control, just as it is in some other animal populations. Man is dependent on the growth of plants, directly as a herbivore and indirectly as a carnivore. Ultimately there must be enough land to support the vast quantities of plants needed to keep the human population alive. This is true even though other forms of fuel are used for social energy requirements. This will remain true unless the human population develops a method for synthesizing food in large quantities without going through the plant or another method to raise more plants. In some parts of the world the food supply *is* a limiting factor for some populations. Yearly many people starve in India and South America. A continued increase in population brought about by a high reproductive potential and a lessening of parasitic and disease control foreshadows more starvation in the future.

In order to stave off mass starvation, the human population must be controlled by man, himself. He has the decision-making ability which caused him to interfere with the population control factors in the first place. He can also decide to control his own population size himself in order to save the environment so that it can continue to support the human population and other populations as well. Other animals cannot make this decision.

Several methods have been proposed and/or tried which could limit the population. These are gericide, infanticide, ritual killing, and birth control. Some or all of these methods are considered unethical or immoral in most cultures. The killing of the aged, gericide, has been an accepted practice in some cultures, for instance in that of the Eskimos. When the old became too weak to hold their own, they were killed or left to die. This approach, however, is rather ineffective in controlling population size since it does not limit the reproducing segment of the population. The information in Table 18-1 indicates that this removal of the elderly would decrease the total population by less than 10%. The population of the twelfth generation would only change from 7680 to 7168 if gericide were practiced. This might give some immediate relief, but it certainly will not stabilize a population.

Since the population-increase rate keeps rising, any method of killing people or transporting people away from the earth would be rapidly outmoded with time unless the facilities for its execution could expand at the same rate to meet the expanding demands. With a current annual increase of about 70,000,000, it is difficult to envision a rocket technology which could get people off the earth fast enough, even if there were some place to which to transport them.

Any method, however, which limits the number of children per family to two will start to stabilize the total population. The population will become constant after about three generations. A population is constant only when the number of births equals the number of deaths. If the average family has been having four children, there are naturally more young people than old people in the population. As long as this is true, the number of deaths will be less than the number of births and the population will grow. As shown in Table 18-2, when the first generation to have only

Generation	Age group				Total population
	Child	25	50	75	
1	16 + 16	4 + 4	2 + 2	1 + 1	46
2	16 + 16	16 + 16	4 + 4	2 + 2	76
3	16 + 16	16 + 16	16 + 16	4 + 4	104
4	16 + 16	16 + 16	16 + 16	16 + 16	128
5	16 + 16	16 + 16	16 + 16	16 + 16	128
6	16 + 16	16 + 16	16 + 16	16 + 16	128

Table 18-2

Stable Population Increase Table for Humans

Assume starting population of two: families with two children; mortality after 75; sex ratio 50:50; brother-sister marriage.

two children per family reaches old age, there are equal numbers of people in all age groups and the population then remains constant.

Birth control is the solution most often proposed to population control. The runaway feedback pattern which the reproductive potential of humans allows could be blocked if the reproductive potential rather than its results is controlled. There are many methods of birth control available. These methods vary in the manner in which they prevent conception and in their effectiveness. The intrauterine device is a device which is inserted into the uterus of the female. It usually prevents pregnancy but it is not known exactly why it does so. There are also physical barriers that can be used by either sex partner to prevent the transmission of semen during intercourse. These methods can be quite dependable if the devices are used consistently. The most dependable method available today is the use of hormone-containing pills that prevent ovulation in the female. The chemical mechanism of action of these pills is as yet unknown, but much research is being conducted to discover this mechanism and to minimize any possible undesirable or harmful side effects.

In the United States birth control methods have been quite widely accepted and the growth rate of the population is well below the world average. Abortion as yet is not completely accepted, but more and more states are gradually "liberalizing" their abortion laws. The threat of overpopulation is forcing man to reevaluate his social customs. The use of birth control measures has been the primary factor that has kept the United States population from increasing even faster than it has. In societies which have different cultural backgrounds, however, birth control methods have not been widely accepted. In such cultures, starvation can be expected to decimate the population periodically unless another method of population control is willfully adopted. Birth control and starvation appear to

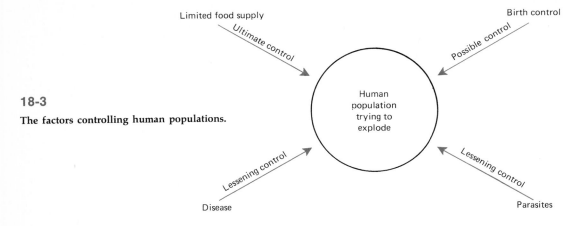

18-3

The factors controlling human populations.

be the most effective factors for limiting human population on earth in the foreseeable future (Fig. 18-3). Whether, then, birth control or "death control" will limit our numbers is uncertain; it seems rather clear that the present rate of increase will not continue indefinitely.

Part VI

Human Ecology and Conservation

Man and Air 19

Introduction

Ecology is the study of the interactions of organisms with one another and their environment. All scientists try to "isolate a system" for study purposes, and ecologists select the ecosystem for study. An ecosystem is any system in nature which includes a specific physical environment and the entire biological community that interacts with it. An important aspect of ecology is the study of how an ecosystem *reacts* to any *change* that occurs within its physical environment or biological community. An ecosystem contains a vast network of interdependencies. Each organism within it depends specifically on other organisms within it as well as on certain elements of its physical environment for survival. An ecologist tries to assay the nature and scope of these many interdependencies. The better he understands these complex relationships, the better he will be able to predict the *outcome* of a particular change that occurs within an ecosystem.

Human ecology focuses attention on the interaction of man with the physical and biological elements of his surroundings. As discussed in Chapter 18, man has been biologically unique in his ability to "control" many aspects of his interaction with his environment. By this "control," however, he has caused many changes in his ecosystem. Since humans inhabit regions all over the world, and man's interaction with his environment is so far-reaching, man's ecosystem is, in fact, the entire biosphere. Since this ecosystem is so immense, man has been able in the past to delude himself that any change in the ecosystem brought about by his

Smog in Denver. (*Courtesy of Dave Cupp.*)

"tampering" with the environment would be too small to be significant. The human population, however, has grown to such a size—by virtue of this "environmental control"—that it is fast becoming obvious that man's effect on the biosphere can no longer be overlooked. All aspects of the biosphere—the biological community, the atmosphere, the waters and the earth materials—have already been considerably changed, and polluted, by man's "progress."

Today pollution looms as a major problem. It will not be sufficiently solved merely by "undoing" the damage that has already been done to the biosphere. The human population is expanding so rapidly that its pollution will certainly expand with it if society continues the current practices which have proven to be unsound ecologically.

With a thorough understanding of the ecology of the biosphere man can foresee the far-reaching effects that his future environmental "tampering" could have. Conservationists consider ecological relationships to help determine the *best* mode for man to interact with the environment in order to minimize any detrimental effects to the ecosystem. Since conservation and human ecology are so closely related, they will be considered together in this text. The principal ways in which man interacts with his environment and the changes he can cause in his ecosystem will be discussed in the next few chapters. By analyzing all the ecological effects caused by the current ways man interacts with his environment, it may be possible to suggest alternate means to the same ends which do not upset the ecological balance of the biosphere.

Air

Air is naturally composed of about 79% nitrogen (N_2), 20% oxygen (O_2), varying amounts of water vapor, and 1% other materials (including CO_2). All these components are gaseous. Air has always carried nongaseous natural contaminants which were temporarily airborne; dust, smoke (from naturally occurring fires), pollen, and decay products are a few such materials. Man's activities, however, have added great quantities of both gaseous and nongaseous contaminants to the atmosphere. In many cases the amount added was large enough to change considerably the physical characteristics of the atmosphere. The magnitude of the "natural contamination" is so small relative to that of "human contamination" that it will be overlooked in this discussion of atmospheric pollution.

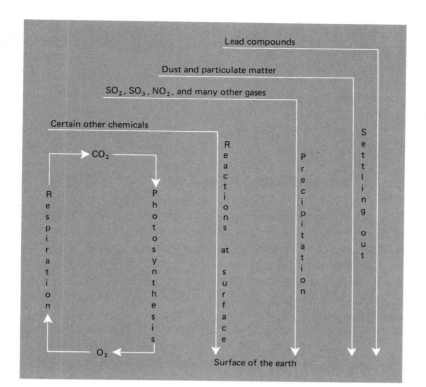

19-1

Air self-purification and regulation. Air will "cleanse" itself rapidly through naturally occurring processes if there is no continuing contamination.

The relatively stable composition of air and its purity tend to be regulated by the natural processes occurring in the atmosphere. For instance, the CO_2/O_2 ratio is maintained by the respiration-photosynthesis reaction cycle. Photosynthetic organisms remove CO_2 from the air, incorporate its carbon content into more complex organic compounds, and release O_2 back to the air. On the other hand, respiring organisms consume this O_2, use it to break down the organic food-fuels produced by the photosynthetic organisms, and regenerate the CO_2. Most unnatural gases in the air usually either dissolve in the earth's waters and precipitate out or react with chemicals at the surface of the earth. Particulate matter which is temporarily airborne eventually settles out. By these processes even air that is heavily contaminated will be eventually "cleaned" of most of its contaminants in a matter of days (Fig. 19-1). It should be noted that, although the air gets cleaned, some of these contaminants which "dissolve" or "settle out" then collect in the waters and on the earth's surface. Materially the earth is indeed a closed system.

Because the atmosphere is "cleaned" naturally, the major problem of atmospheric pollution is not presently one of residual contamination. Even though the air is not constantly contaminated at levels deleterious to organisms, it has been seriously contaminated for brief periods. Air pollution, therefore, must be stopped at its source. Even if this is done, some pollutants, originally introduced into the atmosphere, will remain in the earth's waters and on its surface for years to come. Some pollutants, still in the atmosphere, will persist in global circulation patterns for at least 10 years.

Sources and Effects of Air Contamination

The air gets contaminated with a great many substances from an infinite variety of sources. Automobiles, home furnaces, incinerators, smelters, glass factories, mines, nuclear generating plants, motorcycles, and construction work are just a few examples of everyday factors which contribute to air pollution. Table 19-1 indicates the nature and quantities of contaminants that such sources add to the atmosphere.

Smog The term smog originally referred to the visible atmospheric mixture of *smoke* and *fog*. This term is generally applied now to any visible air contaminant haze, whether it contains fog or not. Nowa-

Table 19-1 Sources and Quantities of Air Contaminants (in millions of tons per year)

Reproduced by permission of the National Tuberculosis and Respiratory Disease Association.

Source	Carbon monoxide	Sulfur oxides	Hydro-carbons	Nitrogen oxides	Particulate matter	Miscel-laneous	Total
Transportation (motor vehicles)	66	1	12	6	1	<1	86
Industry	2	9	4	2	6	2	25
Power plants	1	12	<1	3	3	<1	20
Space heating (homes and factories)	2	3	1	1	1	<1	8
Refuse disposal (mostly incinerators)	1	<1	1	<1	1	<1	4
Totals	72	25	18	12	12	4	143

days smog commonly envelops cities or other large industrial complexes.

The most common component of smog is carbon monoxide, CO. This compound is formed if the burning (oxidation) of carbon-containing fuels (the process which occurs in automobile engines and furnaces) is incomplete. Such a fuel may be completely burned to form only CO_2 and H_2O. A typical reaction of this kind is the combustion of octane, a common fuel.

$$2 \underset{\text{octane}}{C_8H_{18}} + 25\ O_2 \xrightarrow{\text{combustion}} 18\ H_2O + 16\ CO_2$$

In automobile engines or furnaces, however, there may not be enough oxygen present to oxidize these fuels totally. For instance, an automobile requires a 15:1 weight ratio of air:gasoline to oxidize octane completely. This ratio is rarely attained, and incomplete combustion typically results.

$$C_8H_{18} + 11\ O_2 \xrightarrow[\text{combustion}]{\text{incomplete}} 9\ H_2O + 5\ CO_2 + 3\ CO$$

CO is quite toxic to humans. Fortunately CO does get oxidized to CO_2 in the atmosphere. This reaction, however, is slow. CO_2 is relatively harmless to man at low concentrations. Human lungs can contain air with several percent CO_2 without ill effects and, indeed, the lung is especially adapted to eliminate CO_2 from the body of man.

The oxides of sulfur are other major components of smog. Since most organic fuels contain sulfur, these pollutants also result from the combustion of fuels. The sulfur content of fuels can range from about 1 to 5%. When sulfur-containing fuels are burned, the following products appear:

$$\text{Fuel (contains C, H, S)} + O_2 \xrightarrow{\text{combustion}} H_2O + CO_2 + SO_2$$

When SO_2 dissolves in water, even in fog, mist, or rain, it forms a weak solution of sulfurous acid which is corrosive.

Smog also generally contains a wide assortment of hydrocarbons, compounds which contain only carbon and hydrogen. Most fuels are hydrocarbons (for example, octane, C_8H_{18}); therefore hydrocarbons escape into the atmosphere from incomplete combustion of fuels and evaporation of unburned fuels.

There are three kinds of hydrocarbons which are found in smog: paraffins, olefins, and aromatics. Paraffins are compounds which have only single C—C bonds in their molecular structures. They have a general formula $C_nH_{(2n+2)}$ as typified by octane, C_8H_{18}.

$$H-\underset{\underset{H}{|}}{\overset{\overset{H}{|}}{C}}-\underset{\underset{H}{|}}{\overset{\overset{H}{|}}{C}}-\underset{\underset{H}{|}}{\overset{\overset{H}{|}}{C}}-\underset{\underset{H}{|}}{\overset{\overset{H}{|}}{C}}-\underset{\underset{H}{|}}{\overset{\overset{H}{|}}{C}}-\underset{\underset{H}{|}}{\overset{\overset{H}{|}}{C}}-\underset{\underset{H}{|}}{\overset{\overset{H}{|}}{C}}-\underset{\underset{H}{|}}{\overset{\overset{H}{|}}{C}}-H$$

octane

Olefins are compounds whose molecules contain at least one double C=C bond. A simple olefin, produced in the incomplete combustion of gasoline, is ethylene, C_2H_4.

$$\underset{H}{\overset{H}{\diagdown}}C=C\underset{H}{\overset{H}{\diagup}}$$

ethylene

The molecular structure of aromatic compounds contains at least one ring structure of the benzene type. Benzene, C_6H_6, is a compound whose molecular structure has six carbon atoms arranged in a ring.

benzene

Aromatic hydrocarbons are already significant components of smog. A plan to add them to gasoline instead of tetraethyllead is being considered. If this is carried out, aromatics may become an even larger component of smog.

There are several different oxides of nitrogen which are found in the atmosphere. Nitrous oxide (N_2O) occurs naturally in the air as it is formed at the earth's surface. Its concentration, however, is less than 1 part per million (ppm) and it is not harmful at that level. Nitric oxide (NO) is not formed in the atmosphere under natural conditions. It can, however, be formed there at very high temperatures. The high combustion temperatures of gasoline engines and other power plants are able to cause the formation of NO. This compound is relatively harmless, but it oxidizes readily in air to form the toxic nitrogen dioxide, NO_2. Aside from its own toxicity, NO_2 can combine with other atmospheric components under certain conditions to form other toxic substances. It can also combine with water to form a corrosive solution of nitric acid (HNO_3).

Smog also contains particulate matter such as soot and fly ash. These materials are chemically unreactive. Fly ash is the uncombustible inorganic matter which remains after combustion. Soot and other materials are the burnable materials which escaped combustion in an inefficient burner system. These materials settle out of the air as "dustfall." In some highly polluted cities up to 50 tons of dustfall have settled on 1 square mile during a single month. Particulate matter of this sort apparently can serve as a carrier of chemicals and as a site of chemical reactions.

Lead and fluoride compounds are also found in smog. Fluoride compounds are commonly produced in many industrial processes; smelting, ceramic clay product manufacture, glass manufacture, and phosphate fertilizer manufacture are a few such processes. Most fluoride compounds are toxic, but the gaseous hydrogen fluoride, HF, is probably the most toxic. It is highly water soluble and is injurious to plants. Lead compounds get into the atmosphere during the combustion of gasoline to which tetraethyllead has been added as an "antiknock" agent. In sufficient concentration lead is known to be harmful to the human central nervous system. Airborne lead compounds produced by automobiles are being ingested by humans, but it is difficult to determine how damaging to the health this is. Lead has been found to accumulate in the body. Since lead can cause severe damage at certain accumulated levels, a campaign is underway to remove lead compounds from automobile gasoline.

Photochemical Smog

When sunlight interacts with the SO_2, NO_2, O_2, and hydrocarbons present in smog, many chemical reactions can result. These reactions are called photochemical because they require the energy of the sunlight in order to proceed at an appreciable rate. Some of the products of these photochemical reactions are the most noxious air contaminants so far experienced. Such biologically detrimental compounds as ozone, (O_3), SO_3, formaldehyde (H_2CO), acrolein, and peroxyacetyl nitrate (PAN) are all produced in this way (Fig. 19-2).

The mechanisms of these photochemical reactions are not well understood. Apparently the NO_2 in the atmosphere is photoreactive and reacts to form NO and the "reactive intermediate," the oxygen atom, O:

$$NO_2 \xrightarrow{sunlight} NO + O$$

19-2

A summary of important results of photochemical reactions leading to the production of some of the most noxious chemical contents of urban smog. PAN, formaldehyde, and acrolein in moderate concentration are very irritating to humans and toxic for plants. Ozone is very reactive and helps form such compounds as SO_3 which easily takes on water to form sulfuric acid (H_2SO_4).

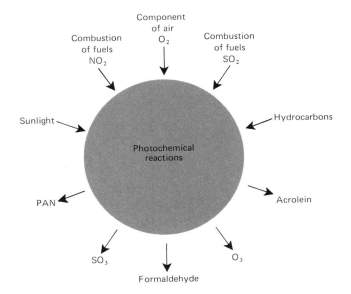

This intermediate can react with other components of smog such as O_2 and SO_2 to form ozone and sulfur trioxide, respectively:

$$O + O_2 \longrightarrow O_3$$
$$O + SO_2 \longrightarrow SO_3$$

Ozone is also photoreactive and reverts to O_2 and the oxygen atom:

$$O_3 \xrightarrow{\text{sunlight}} O_2 + O$$

The presence of ozone and sunlight, then, can also bring about the formation of SO_3 from SO_2. In addition formaldehyde and acrolein are formed when O_3 photochemically reacts with various terminal olefins:

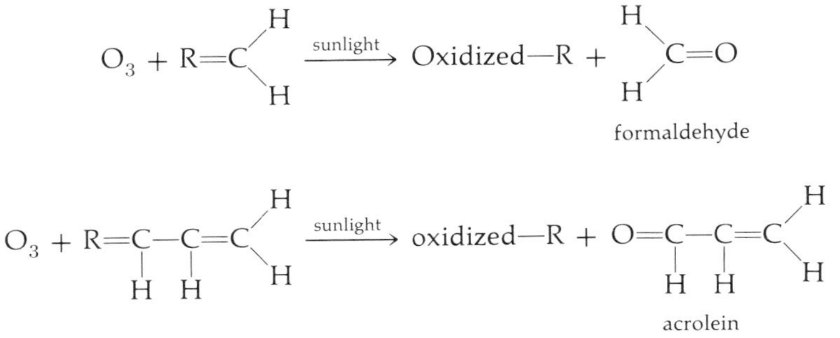

$$O_3 + R{=}C\genfrac{}{}{0pt}{}{H}{H} \xrightarrow{\text{sunlight}} \text{Oxidized}{-}R + \genfrac{}{}{0pt}{}{H}{H}{>}C{=}O$$

formaldehyde

$$O_3 + R{=}\underset{H}{\overset{}{C}}{-}\underset{H}{\overset{}{C}}{-}C\genfrac{}{}{0pt}{}{H}{H} \xrightarrow{\text{sunlight}} \text{oxidized}{-}R + O{=}\underset{H}{\overset{}{C}}{-}\underset{H}{\overset{}{C}}{=}C\genfrac{}{}{0pt}{}{H}{H}$$

acrolein

Similar, but more complicated photochemical reactions result in the formation of PAN.

Human Ecology and Conservation

$$\text{H}-\underset{\underset{\text{H}}{|}}{\overset{\overset{\text{H}}{|}}{\text{C}}}-\overset{\overset{\text{O}}{\|}}{\text{C}}-\text{O}-\text{O}-\text{N}\overset{\nearrow \text{O}}{\searrow \text{O}}$$

peroxyacetyl nitrate (PAN)

All these and many other products of the photochemical reactions that occur in smog have unpleasant and/or harmful effects on animals, plants, and materials such as paint and rubber. This photochemical interaction of the contaminants which collect in smog makes the smog phenomenon much more deleterious than was originally presumed.

Effects of Smog on Humans Polluted air is a very complex mixture which changes composition daily. It is therefore difficult to *prove* that a specific biological or physical change in the biosphere results from air pollution. It is even more difficult to identify a specific contaminant as the cause of a specific effect. A great deal of evidence, however, is emerging that indicates that the health of humans subjected to a low level of air contamination over long periods of time is indeed adversely affected.

A general correlation between air pollution and respiratory problems has been noted. For example, in the United States the percentage of deaths resulting from lung cancer and emphysema (a lung disease) are twice as high in large metropolitan areas as they are in rural areas. Furthermore, the percentages of people dying from either of these two diseases are increasing, paralleling increases in air pollution. In addition, autopsies on the lungs of 300 smokers from the heavily polluted St. Louis, Missouri, area indicated four times the severe emphysema as in the lungs of 300 smokers from the less polluted area, Winnipeg, Canada. A 6-year study of the absenteeism of British postmen due to bronchitis showed that postmen working in heavily polluted areas were absent three times as often as the postmen working in less polluted areas.

The oxides of sulfur, SO_2 and SO_3, in conjunction with particulate matter are under suspicion as being air contaminants which contribute to ill health. Although the evidence implicating these compounds is inconclusive, a campaign is underway to reduce the sulfur content of fuels and to remove sulfur oxides which escape from smokestacks. It may be, however, that the other smog components and their photochemical products are the causes of ill health as well as, or instead of, the sulfur oxides. The photochemical products formaldehyde, acrolein, and PAN are known to be eye irritants. They might also interfere with the respiratory system.

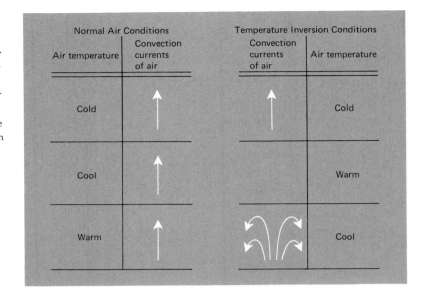

19-3

Airflow patterns for normal air conditions (left) and temperature inversion air conditions (right). The lack of air convection movements under temperature inversion conditions explains the progressive build up of air contamination near the ground level.

There is no evidence which suggests that any of the smog contaminants can be eliminated from suspicion as biological irritants.

An atmospheric condition called a temperature inversion allows smog to build up to very high and dangerous levels (Fig. 19-3). During the day the air closest to the earth's surface is generally warmer than the air farther away from the surface. Since warm air is less dense (lighter) than cold air, it rises into the cold denser air above it. By this natural process the contaminated air at the earth's surface rises into the colder upper air and is dispersed. During a temperature inversion, however, the layer of air at the earth's surface is colder than that farther away. This air at the surface, then, does not naturally rise. In fact, the warm air layer above it acts like a lid. Pollutants therefore accumulate in the air at the earth's surface and are not dispersed until the inversion finally breaks.

Temperature inversions of this sort have actually caused lethal atmospheric conditions to develop. Such conditions led to 60 deaths in the Meuse Valley of Belgium in 1930, 20 deaths in Donora, Pennsylvania, in 1948, 4000 deaths in London, England, in 1952, and 168 deaths in New York City in 1966. Smog build ups are not always so drastic, but they do seem to have an effect. The number of hospital admissions of cases with allergic disorders, respiratory infections or diseases, and heart or vascular diseases for one city hospital for over 200 consecutive days correlated quite well with the air pollution daily levels for the period.

Effects of Smog on Vegetation Some of the components of smog have been shown to be toxic to plants. Some plants are so sensitive to these materials that they exhibit reproducible symptoms of effect even after exposure to a very low concentration (a few ppm) of contaminant for only a few hours. For example, ethylene causes damage to orchid flowers at 0.05 ppm for 6 hours and causes abnormal bud opening of carnations at 0.1 ppm for 6 hours. Fluorides have been found to accumulate in sensitive plants and cause leaf damage at atmospheric concentrations as low as 0.0005 ppm. Leaf damage to sensitive plants is caused by NO_2 at 3 ppm. Ozone can severely damage bean and tomato plants at 0.4 ppm for 2 hours and damage ponderosa pines at levels of 0.5 ppm for 7–10 day periods. A concentration of PAN of 0.1 ppm for 5 hours damages severely bean, petunia, and tomato plants. Plants sensitive to SO_2 accumulate sulfates in their leaves. This can ultimately lead to leaf damage. In general the experimental levels of contamination are higher than typical present-day levels of air contaminants. The contaminant levels are generally never more than 0.05–0.2 ppm for NO_2, 0.05–0.2 ppm for O_3, 0.02–0.2 for PAN, up to 0.5 ppm for ethylene, and up to 0.2 ppm for SO_2. Concentrations of these compounds, however, do sometimes reach harmful levels in smog conditions, and in each case the toxic result of the compound is evident.

The chronic effects of smog on plants are also difficult to identify and prove, as in the case with humans. Evidence, however, does indicate that plants are significantly effected by long-term exposure to low levels of air pollution.

Some studies of various plants in the heavy smog area near Los Angeles show effects of this pollution. The Los Angeles smog appears to be killing the ponderosa pines of the San Bernadino National Forest at a rate of 3% a year. The photosynthesis of these pines apparently is reduced by 10%, 70%, and 85% by exposure to ozone at the respective concentrations of 0.15, 0.30, and 0.45 ppm for 30-day periods. Lemon and navel orange trees grown in filtered air (relatively clean) in the Los Angeles area yielded more fruit and registered more water uptake and photosynthesis than trees grown in unfiltered (contaminated) air. Zinfandel grape plants grown in filtered air in this area yielded more and sweeter grapes than plants grown in unfiltered air. The thickness and chlorophyll content of the leaves were also greater for the plants grown in the smog-free environment.

These examples indicate the kind of damage long-term, low-level contamination has been shown to cause plants. Undoubtedly, less evident damage is also occurring. Decreased food production due to air pollution can certainly undermine the attempts to increase food production to feed an increasing population.

Effects of Smog on Materials Air contamination causes material damage as well as biological damage. All external surfaces, be they on statues, buildings, cars, bridges, or any other, are corroded by air contaminated with acids and other corrosive substances. Ozone causes rubber and paint to dry and crack. Dustfall soils everything on which it lands. Maintenance and cleaning costs resulting from this pollution damage are estimated to be $65 per person per year for the entire United States population and as high as $75 in highly polluted areas. These costs of air pollution damage have great bearing on the economics of air clean up.

Nuclear Radiation Radioactive materials have unstable nuclei which give off one of three kinds of high energy radiation: alpha radiation (He^{2+} ions, particles with two protons and two neutrons), beta radiation (electrons), and gamma radiation (waves similar to x-rays). By this radiation these materials are ultimately transformed to stable non-radioactive materials. However, this transformation occurs very slowly for many such materials. For instance, the time it takes for half the radioactive material to decompose is 28 years for strontium-90, 30 years for cesium-137, 1600 years for radium-226, and 24,000 years for plutonium-239. Even a very slow release of these materials into the environment will result in a serious build up of these contaminants.

The energy of nuclear radiation, however, is much higher than that of most chemical bonds. It is, therefore, capable of causing changes in compounds which are normally considered stable. Radiation can change the structure of important cellular compounds such as DNA. If the DNA of reproductive cells is altered, random mutation can occur which may be lethal to offspring. Radiation may also change normal cells into cancerous ones.

There are many natural occurring radioactive compounds, so the biological community has been continuously subjected to a low level of radiation, called background radiation. Presumably this level causes few major ill effects on biological systems. However, through man's recent capability to "harness" nuclear power, the level of radiation in the environment is rising. During both the development and testing of nuclear weapons and electrical generation from nuclear power, many radioactive substances have accidently been introduced into the environment. The surface testing of atomic warheads has littered the world with strontium-90. An accidental fire at the Rocky Flats Plant in Colorado released some plutonium-239 into the air. The flaming of natural gas released by a nuclear explosion in the Rulison Project in Colorado released hydrogen-3 (tritium) into the air.

It has been estimated that man has introduced only 1% more radioactive materials into the environment than was naturally present. Yet the cumulative nature and the potential danger to organisms of radioactive materials make the safety of introducing any additional radioactivity questionable. However, the percentage of electricity generated by nuclear power plants is predicted to increase from 1% to 30% by 1980 and to 50% by 2000. The possible long-term danger of radioactive contamination even at low levels makes the maximum possible control of radioactive leaks seem imperative. If such intensive controls are proven unnecessary in the future, it will be far easier to relax the controls than to extract the dangerous and irretrievable radioactive materials from the environment.

Noise

Noise is unwelcome sound. Some sounds such as the whistle of wind through the trees, the lapping of waves on a beach, or the chirping of a bird announcing spring would probably be considered pleasant to many. Other sounds, however, such as that of an air hammer attacking cement, a jet taking off, or the impatient honking in a traffic jam are not as euphonious. These are considered noise by most people. As the population mounts, so does the amount of unwanted sound. In more populated areas noise is gradually becoming an important component of air pollution.

Continued exposure to certain levels of noise apparently leads to progressive hearing loss. The hearing loss for people with jobs in noisier areas is apparently greater than for those employed in quieter areas. For instance, factory workers have more hearing loss than office workers. In general, hearing loss has increased among workmen paralleling the increase in industrialization. Noise also appears to elicit systematic stress-type reactions in some people; blood vessel constriction, muscle constriction, and adrenalin release are three such stress-type reactions. Noise also has the more obvious effects of interfering with sleep, rest, and relaxation.

The Future

City dwellers are becoming increasingly aware that something must be done about air and noise pollution. Solutions to this problem, however, are not simple. Much of the pollution has arisen from the production of goods and services which people now take for granted. This production cannot be stopped simply. Instead, new methods of production which are more ecologically sound must be developed.

The Ecological Debt   The total cost of production of goods and services must be paid by the consumer. In the past part of these production costs were paid by the depreciation of the environment. This has produced an "ecological debt"—an accumulated waste and pollution of the environment that is now barely tolerable. In the future the cost of any product or service must also include the cost of air pollution suppression. The power user will have to pay for electricity generated by a process which does not contaminate the atmosphere. The commuter will have to pay for clean transportation.

The products or services offered by nonpolluting producers will necessarily be more expensive. The nonpollutors stand to be driven out of business by competing firms who can offer similar goods at lower costs because they do not pay the cost of pollution suppression. *Nationwide* controls on the allowable pollution levels will have to be extended to avoid such a catastrophe brought about by economics. State control would be insufficient as it would only drive pollutors to states with less strict pollution laws.

As the population density increases, the recent parallel trend has been a decrease in the freedom for the individuals in the population. The mountain man in the Old West, for instance, had individual freedoms that are unknown to the present-day city dweller. Even a small town citizen is bombarded with permits, licenses, traffic signs, parking regulations, and zoning laws. A big city dweller is subject to even more restrictions. Even more regulations may be imposed on society unless individuals themselves decide that the environment must be saved. They must start to weigh heavily the depreciation of the environment and its effects when evaluating the worth and convenience of a service or product which pollutes the environment. For instance, if a car commuter took into account the "social" costs of running his car—air pollution, crop losses, road maintenance, parking space, dwindling petroleum and mineral resources, law enforcement personnel, traffic installations, noise, and so forth—he might decide that clean mass transportation is more "economical." Otherwise mandatory use of rapid transit, bans of automobiles from core cities, curfews on power lawnmowers and motorcycles, gasoline rationing, and other such regulations might well be necessary.

Transportation and Smog

Both technological and sociological advances are needed to eliminate the air pollution from all modes of transportation used in the United States. The perfection of an economical, mass producible, and nonpolluting car would help the smog problem considerably without disrupting very many patterns of living. It is speculated, however, that such a car is still 5–20 years in the future. The internal combustion engine must be cleaned up as far as is possible in the meantime. The design of a low-pollution rapid-transit system would also considerably alleviate the city smog problem, especially if it caused a decrease in automobile commuting.

Research

Although billions of dollars are spent each year on weapons, rockets, tobacco, liquor, and drugs, only millions are spent on environmental research. Solutions to air pollution might well be found if more researchers devoted their energies to environmental problems and if private, state, and federal agencies supported such research as well as they currently support many other fields.

References and Recommended Readings

AARONSON, TERRI. "Tempest Over a Teapot," *Environment,* **11** (8), 22–27 (1969).

A Primer on Air Pollution. 2nd ed., Mobil Oil Corporation, 150 East 42nd Street, New York 10017, 1970.

"Air Pollution." *Time Mag.* editorial, January 27, 1967 (pp. 48–52).

Air Pollution Primer. National Tuberculosis and Respiratory Disease Association, 1740 Broadway, New York 10019, 1969.

ALLEN, S. W., and J. W. LEONARD. *Conserving Natural Resources,* McGraw-Hill Book Company, New York, 1966.

BENARDE, MELVIN A. *Our Precarious Habitat,* W. W. Norton, New York, 1970.

CADLE, R. D., and E. R. ALLEN. "Atmosphere Photochemistry," *Science,* **167,** 243–249 (1970).

"City v. Forest." *Time Mag.* editorial, April 13, 1970 (pp. 49).

Comparative Emissions from Some Leaded and Prototype Lead-Free Automobile Fuels. Report of Investigations no. 7390, Bureau of Mines, United States Department of the Interior.

Control of Automobile Emissions. Technical Information Service, Public Relations Staff, Ford Motor Company, Dearborn, Michigan, 1969.

DARLEY, E. F., C. W. NICHOLS, J. T. MIDDLETON. "Identification of Air Pollution Damage to Agricultural Crops," *The Bulletin* (Department of Agriculture, State of California) **55**, 11–19 (1966).

EHRLICH, P. R., and A. H. EHRLICH. *Population, Resources, Environment*. W. H. Freeman, San Francisco, Calif., 1970.

GOFMAN, J. W., and ARTHUR TAMPLIN. "Radiation: The Invisible Casualties," *Environment* **12** (3), 11–19, 49 (1970).

Keep It Clean: Highlights of Bethlehem's Pollution Control Program. Bethlehem Steel Corporation, Bethlehem, Pennsylvania, 18016, 1970.

LEAR, JOHN. "Green Light for the Smogless Car," *Saturday Rev.*, December 6, 1969 (pp. 81–86).

LEAR, JOHN. "A Progress Report on Smogless Motoring," *Saturday Rev.*, August 1, 1970 (pp. 44–45).

MARTELL, E. A., et al. "Fire Damage," *Environment*, **12** (4), 14–21 (1970).

MILLER, P. R., et al. "Ozone Dosage Response of Ponderosa Pine Seedlings," *APCA J.*, **19**, 435–438 (1969).

"Pollution Price Tag: 71 Billion Dollars." *U. S. News World Rep.*, August 17, 1970 (pp. 38–42).

TAYLOR, O. CLIFTON. "Effects of Oxidant Air Pollutants." *J. Occup. Med.*, **10**, 485–492 (1968).

TAYLOR, O. CLIFTON. "Importance of Peroxyacetyl Nitrate (PAN) as a Phytotoxic Air Pollutant," *APCA J.*, **19**: 347–351 (1969).

TAYLOR, O. CLIFTON. "Agriculture and Air Pollution," *Calif. Air Environ.*, **1**, 1–3 (1970).

THOMPSON, C. R., et al. "Effects of Air Pollutants on Apparent Photosynthesis and Water Use by Citrus Trees," *Environ. Sci. and Technol.*, **1**, 644–640 (1967).

THOMPSON, C. R., E. HENSEL, and G. KATS. "Effects of Photochemical Air Pollutants on Zinfandel Grapes," *Hort. Sci.*, **4**, 222–224 (1969).

THOMPSON, C. R., and O. C. TAYLOR. "Effects of Air Pollutants on Growth, Leaf Drop, Fruit Drop, and Yield of Citrus Trees," *Environ. Sci. Technol.*, **3**, 934–940 (1969).

THOMPSON, C. R. "Effects of Air Pollutants in the Los Angeles Basin on Citrus," *Proc. 1st Int. Citrus Symp.*, **2**, 705–709 (1969).

Man and Water 20

Water is, and always has been, an integral part of human existence. Although each day a man's body needs only about 1 gal of water, approximately 3500 gal of water are required to grow the food he eats that day. In addition, man puts water to various "social" and "technological" uses which consume about 1500 gal per man per day. The "social" uses include washing, laundry, sewage treatment, lawn watering, and so on. The "technological" water requirements are about 10 times as great as these "social" ones. For example, 250 gal of water are needed in the manufacture of only 1 lb of newsprint, 60 gal/lb of aluminum, and 14 gal/lb of steel. The total amount of water each man in a technological society requires, then, comes to approximately 5000 gal/day.

The water on the surface of the earth is constantly evaporating into the air. When the air gets supersaturated and can hold no more water vapor, this water returns to the earth in some form of precipitation—rain, sleet, snow, and so forth. This pattern of water movement is called the hydrological cycle (Fig. 20-1). The water supplied to the United States for purposes other than growing food by this cycle averages about 10,000 gal per person per day. (This rate is based on the assumptions that the continental United States is about 2 billion acres, that the average annual precipitation is 30 in., and that only 5% of the precipitation collects in public fresh water supplies.) This supply certainly exceeds the per capita biological, "social," and "technological" needs of only 1500 gal/day, yet water for these purposes is still in short supply in the United States. Why? For one thing water does not always collect where it is needed. The more deplorable reason, however, is that the water collected is too contaminated to be used. Although the water falling

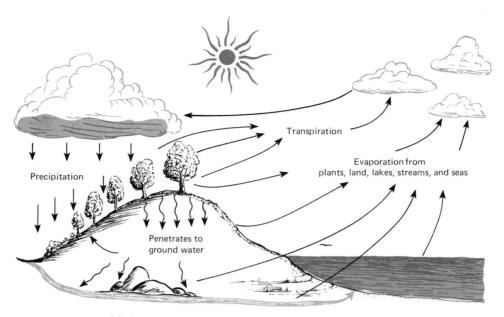

20-1 **The hydrological cycle of evaporation and precipitation.** All but the deep groundwater is carried through this cycle regularly, providing a constantly renewing supply of pure water to the land.

to the earth is relatively pure, once it hits any of the bodies of water— lakes, rivers, streams, oceans, and so on—on the earth, it gets mixed with the contaminants already present in these bodies. Polluted water cannot fill any of the needs which require pure water. The problem is one of purity, then, not of quantity.

Biochemical Oxygen Demand

The earth's waters contain dissolved oxygen. This oxygen comes from the air or from the photosynthesis of the aquatic plant life. Many organisms within the waters require this oxygen for the respiratory break down of the organic compounds which serve as their food. The foods available to them in the waters are the other aquatic plants and animals and the naturally occurring organic wastes from other organisms, both aquatic and terrestrial. The overall reaction for this aerobic break down of organic compounds depends upon the nature of the organic food. Fats and carbohydrates, for example, are broken down to CO_2 and H_2O:

$$\underset{\text{Contains C, H, O}}{\text{Fat or carbohydrate}} + \text{abundant } O_2 \longrightarrow H_2O + CO_2 + \text{energy}$$

[256] Human Ecology and Conservation

Since proteins are chemically more complex and commonly contain nitrogen, sulfur, and phosphorus, a wider variety of products are formed from their break down:

Protein + abundant $O_2 \longrightarrow$
contains C, H, O, N, S, P

$$CO_2 + H_2O + \underset{\text{nitrate ion}}{NO_3^-} + \underset{\text{sulfate ion}}{SO_4^{2-}} + \underset{\text{phosphate ion}}{PO_4^{3-}}$$

All these break-down products, however, are simple inorganic compounds which are excreted back into the waters.

Some organisms, mostly bacteria, can break down organic nutrients anaerobically. The final products from this anaerobic break down, however, are not the completely oxidized CO_2 and H_2O but rather products of only partial break down, such as CH_3CH_2OH (ethyl alcohol), H_2S (hydrogen sulfide), CH_4, NH_3, and so forth.

The oxygen requirement of organisms for the aerobic break down of food in water, into the simple inorganic compounds illustrated in the equations, is called the biochemical oxygen demand, BOD. This is the amount of O_2 withdrawn from the water as aquatic organisms break down their food.

Eutrophication

Eutrophication is the ecological aging of freshwater systems. It is a natural phenomenon that is speeded up by the addition of nutrients as illustrated in the following paragraphs. The main nutrients added to such water systems by man, of course are the organic wastes and inorganic materials of his sewage. Since there are any number of aquatic organisms which can feed on the wastes of other organisms, including man's, this "seems" a natural way to get rid of the wastes in his sewage. The more waste is added to the earth's waters, however, the greater will be the populations of the organisms that can utilize this waste as foodstuff. As the populations of these forms increase, so do the populations of the larger forms which feed on these smaller forms.

In addition, the inorganic materials, particularly nitrates and phosphates added to the water either directly with the sewage or through the biological break down of organic wastes, increase the capacity of the water to support plant life. The increased nitrate and phosphate content of the water increases the growth of water plants just as nitrate and phosphate fertilizers enhance terrestrial plant growth. Enhanced aquatic plant life also allows for larger animal populations in the water. The plants are not only a food source; they also supply more oxygen to the water as they photosynthesize. Clearly the entire aquatic ecosystem is greatly affected by man's addition of nutrient materials.

The aquatic ecological alterations ultimately brought about by humans, however, are not always desirable to them. Although accelerated eutrophication may produce waters well stocked with fish for food and sport, it can also result in water choked with weeds or green with algae and in beaches smelly with windrows of decaying plants.

More serious conditions can also result as eutrophication continues. The many organisms that such high nutrient water can support will have a high oxygen demand. If the oxygen demand ever exceeds the oxygen content of the waters, the balance of the aquatic life in these waters can be seriously altered. At worst the oxygen supply may be depleted to levels at which many of the larger organisms with high O_2 demands die and only the anaerobic bacteria can thrive. At this point the aquatic ecosystem collapses and the waters often become smelly from the anaerobic breakdown products, CH_3CH_2OH, H_2S, NH_3, and so on.

Chemical Contamination

As man has become more and more industrialized, more and more chemicals have found their way into the earth's waters. The major chemical contaminants and their sources are listed in Table 20-1. Hopefully, it is finally being realized that these contaminants are potentially dangerous in anything but trace amounts.

The effects of these contaminants are varied and depend on the nature of the contaminant. The common salts such as NaCl (sodium chloride) and $CaCl_2$ (calcium chloride) can upset the osmotic balance of the water system. Cyanides, mercury compounds, and phenols, however, are toxic to many aquatic organisms. Nitrates and phosphates, as mentioned earlier, speed up eutrophication. Nitrates may also be toxic to biological systems. They can, under certain conditions, be converted to poisonous nitrites in the digestive tracts of humans, particularly in children. Acids and bases are not only toxic, they also upset the pH balance of aquatic ecosystems. Detergents interfere with the action of sewage plants and are detrimental to some aquatic life. Oil coats the surface of water. This interferes with the water's oxygen uptake and coats the feathers

Table 20-1

Chemical Contaminants of Water with Common Sources

Chemical	Common source
Mercury compounds	Paper, plastic, chemical production, some fungicides
Nitrates	Leaching of fertilizers, municipal sewage, air pollution
Phosphates	Leaching of fertilizers, municipal sewage, break down of detergents
Cyanides	Steel manufacture, metal finishing, chemicals production
NaCl, $CaCl_2$, MgCl (salts)	Chemical plants, oil fields
Phenol	Steel manufacture
Detergents	Municipalities
Oil	Spills, used oil, manufacture, leakage
Acids	Mine drainage, steel manufacture, chemicals production
DDT, dieldrin, chlordane, etc.	Insect control
2-4-D, 2-4-5-T	Weed control
Polychlorinated biphenyls	Industrial wastes

of aquatic birds. Insecticides seem to have serious effects on aquatic plant and animal life. DDT is known to be injurious to fish and to the reproductive success of some birds. In addition, many contaminants give water a bad odor or taste and thereby interfere with its use.

Persistent Contaminants

Some contaminants, inorganic and organic, are not quickly broken down by interaction with either the physical or biological environment. DDT and other similar insecticides appear to have a long lifetime on land. It has been estimated that DDT has a half-life of 10 years in the environment. It would take 70 years, therefore, to reduce the DDT concentration in the world to less than 1% of its present concentration **even if all use was immediately stopped.** Exceptionally stable compounds such as this tend to accumulate in the earth's waters and therefore are spread great distances. DDT, for instance, has been found in the bodies of Eskimos in the Arctic and in penguins and seals in the Antarctic. Presumably, almost everything in between is contaminated with DDT and other persistent pesticides as well. Some inorganic mercury compounds, introduced into the environment in fungicides and industrial wastes, are reaching concentration levels which are dangerous for some organisms. Other materials currently in use may also emerge as persistent and dangerous contaminants in the future.

Biological Concentration

Very low levels of contaminants in the waters initially considered "safe" can turn out to be dangerous to certain organisms because of the phenomenon called "biological concentration." The bodies of many organisms are able to "tolerate" contaminants they ingest during feeding. Once ingested, these trace contaminants may not be excreted. As a result the amount of these contaminants can build up within organisms to concentrations far greater than those of the surrounding environment. For example, mercury compounds which had been released into the Minamata Bay in Japan by a nearby industrial plant were picked up by the shellfish who concentrated it greatly within their bodies. Similarly, oysters in the Gulf of Mexico were found to be able to concentrate DDT in their bodies to a level that was 70,000 times that of the surrounding water.

Since one organism feeds on many other organisms for food during his lifetime, biological concentration of a particular contaminant can reach enormous proportions once a contaminant enters a food chain. This point is well made by a chemical analysis of the aquatic life of a lake in California well after it had been sprayed with an insecticide called DDD. The levels of DDD found in the biological community are given in Table 20-2.

In fact, the DDD levels attained by the grebes was toxic to the species and many grebes died as a result. In many cases a species can quickly accumulate more than the species can "tolerate" from its food chain even though the level of that contaminant in his environment is "safe." The people who ate the mercury-infested shellfish from the Japanese bay were severely poisoned by that level

Table 20-2

Biological Concentration of DDD in One Aquatic Ecosystem

Life Form	DDD concentration by life forms
Microscopic life	250×
Frogs	2,000×
Sunfish	12,000×
Grebes	80,000×

of mercury. A biological community, then, can be severely disrupted by water pollution that may originally be considered very low.

Water Treatment

Water can be treated either before utilization by society or afterwards. The treatment of water after use minimizes the amount of water contamination and, therefore, the amount of pretreatment necessary, keeps the environment clean, and does not accelerate eutrophication to undesirable levels. More communities must decide to treat their "used" water if natural plant and animal communities and relatively uncontaminated waters are to be preserved.

Sewage Treatment

Common sewage, that is "used" water, is a very complex mixture containing everything from sticks and paper to submicroscopic dissolved compounds, organic and inorganic. The most common treatment to "purify" this mixture in practice today consists of a sequence of steps in which each subsequent step yields a higher purity of water (Fig. 20-2).

Sewage is first screened to remove the coarse materials. It then passes into a grit chamber where the fine, heavy particles which eluded the screen settle out. The sewage then flows into a sedimentation tank where it stays for $1/2$ to $1\frac{1}{2}$ hours. During this time even finer particles have a chance to settle to the bottom. If no further treatment is to follow this primary treatment, the effluent from this tank is chlorinated and returned to the environment. This effluent, however, still contains very fine particles, colloids, and dissolved materials. If it is released into the environment "as is" it can cause serious problems unless it is highly diluted. The sludge removed from the water during primary treatment is often used as land fill.

This effluent from primary treatment should be treated further

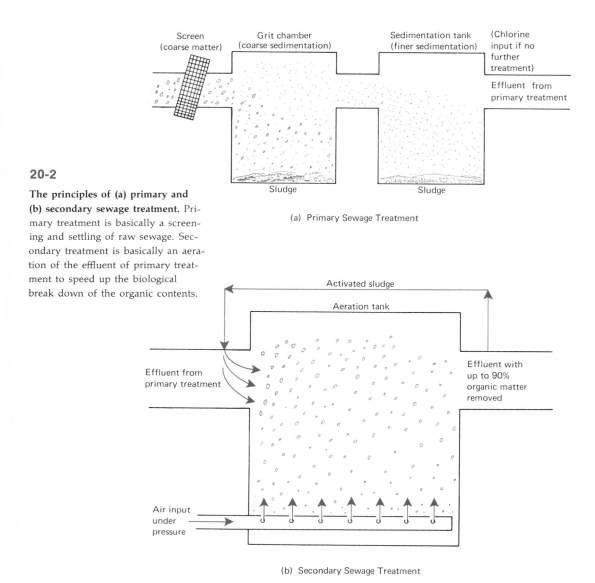

20-2

The principles of (a) primary and (b) secondary sewage treatment. Primary treatment is basically a screening and settling of raw sewage. Secondary treatment is basically an aeration of the effluent of primary treatment to speed up the biological break down of the organic contents.

to speed up the biological break down of the organic materials still present in it. If the sewage effluents can be rid of these organic compounds *before* they are returned to the earth's waters, they will not turn these waters into the rich nutrient soups which can seriously upset the aquatic ecological balance. This accelerated biological break down is achieved by either trickling the sewage over a rock bed and thereby exposing it to aerobic attack by bacteria and other organisms or by running it through an aeration tank where this bacterial attack can also occur (Fig. 20-2b). If allowed to proceed long enough, treatment of this nature can remove up to 90% of the organic materials present in the primary effluent. This secondary effluent still contains some organic compounds, salts, and all the break-down products of the organic wastes including phosphates and nitrates. This effluent is not fit for direct reuse and is still contaminating to the environmental waters.

In recent years methods of tertiary treatment have been developed to purify sewage further (Fig. 20-3). In some cases chemicals such as lime, alum, or synthetic organic compounds are added to the secondary effluent. These compounds will coagulate with the

20-3 Advanced methods of treating sewage that has already received primary and secondary treatment.

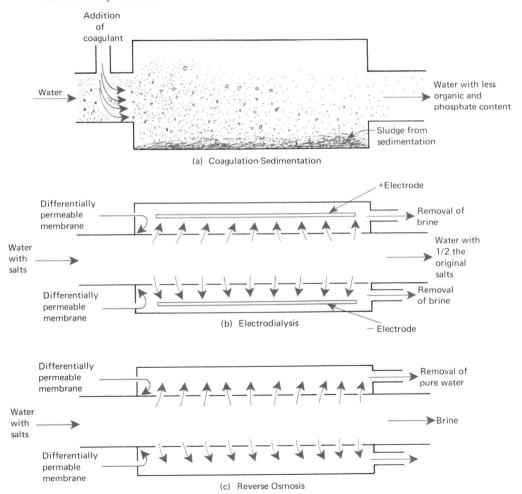

phosphates and organic materials still present and form larger, insoluble particles. These large particles then settle out in a sedimentation tank (Fig. 20-3a). Any organic material still remaining in the effluent after this treatment is removed when the effluent is filtered through activated carbon. The water emerging from this coagulation-sedimentation treatment still carries some inorganic contaminants. It is, however, acceptable for general irrigation, industry, and water sports.

Inorganic compounds can be removed by electrodialysis treatment (Fig. 20-3b). This process reduces the inorganic contaminants by at least 1/2 by attracting the charged ions of the dissolved inorganic electrolytes through semipermeable membranes toward charged electrodes. Another process removes inorganic materials from water by reverse osmosis (Fig. 20-3c). If enough pressure is applied to contaminated water within a semipermeable membrane surrounded by pure water, the osmotic pressure can be overcome and pure water will flow from the solution through the membrane. This process yields water that is drinkable.

Industrial Waste Treatment

Industrial sewage contains large volumes of wastes which are predominantly inorganic. Since inorganic compounds cannot readily be broken down into simpler and more harmless compounds, industrial waste treatment faces different problems than those solved by municipal waste treatment. Industrial wastes are usually decreased by changing an industrial process to a "cleaner" one, by recycling the chemicals involved, or by reclaiming useful products from what was formerly considered "waste." In some cases industry has developed ways to increase the recovery rate of former "wastes" for recycling or reclamation even at a profit.

The cheese and dairy industry produces annually over a billion pounds of whey solids in dilute suspensions. About 70% of these solids were formerly wasted. Recently, however, processing techniques have been developed which can convert these former pollutants into animal feeds as well as into a source of lactose and high-grade protein for human food.

Very toxic waste products such as chromic acid, nickel sulfate, and cyanides are produced by the metal plating industry during copper, zinc, and cadmium plating. The recovery of these electrolytes by electrodialysis promises a good financial return as well as cleaner water. It has been proposed that electrodialysis and evaporation could make possible the recovery of some of the chemicals used in the pulping process as well as some of the organic wastes formed in the process. The former can be recycled in the pulp process and the latter can be converted to profitable products such as binders, glue, and even nutrients.

In general, the technology to correct water pollution problems is well developed. There must be sufficient social or economic motivation to implement these relatively costly means of reducing water pollution.

Purification of Water

Water purification is the term used for the treatment of water prior to use. The type of purification procedure used, of course, depends on the original purity of the water and its intended use. Extremely high-quality water may not even need chlorination to be suitable to drink. The basic series of steps in water purification, however, are aeration, coagulation-sedimentation, filtration through a medium such as sand and gravel, and chlorination. The use of these steps varies according to the starting condition of the water.

The technology to purify water that has a high concentration of dissolved salts has been advancing in recent years. The techniques of reverse osmosis, electrodialysis, and high-volume distillation have been directed toward the production of pure water from the sea or brackish water sources. In the past the barrier to such purification procedures was cost. The current freshwater sources yield water at a cost of less than $0.50 for 1000 gal. The same volume of desalted water, however, costs approximately $0.50–1.00. As traditional freshwater sources require more purification treatment and desalinization techniques improve, these costs are approaching one another. Because brackish water has a lower salt concentration than seawater and may be available in the immediate area in which pure water is needed, its desalinization is rapidly becoming economically feasible. A major problem to desalinization techniques is the disposal of the brine or salt recovered. Approximately 80 lb of salt are produced from every 1% salt content of 1000 gal.

Water Recycling

The direct reuse of water after it has been purified and before it is released into the environment is called the recycling of water. Water recycling requires that treated waste water be pure enough to be used either directly or with only a limited amount of dilution. The waste water treatments currently practiced do not produce water that can be recycled directly. Waste water would have to be given not only primary and secondary treatment but also coagulation-sedimentation and carbon adsorption treatment to attain the purity necessary for most uses. This type of treatment is rare.

Evidently, however, society is evolving toward water recycling (Fig. 20-4). The dense populations in certain parts of the United States make supplying water to these areas impossible on a one-use basis. The water that is currently being reused in these areas is treated after use, released into the environmental waters for major dilution, and later taken in and purified for reuse. The older methods of simply dumping wastes into waterways with little or

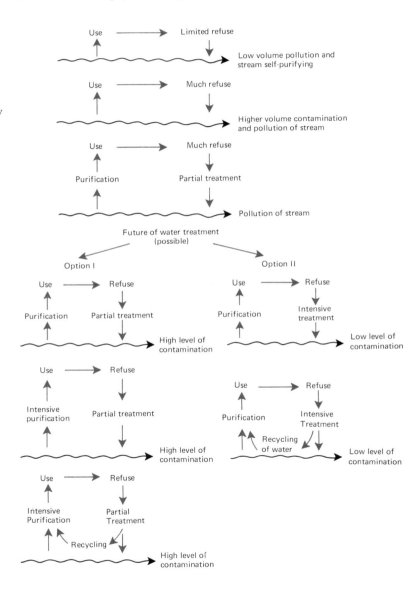

20-4

The history and possible projection of water treatment. Two possible paths for the evolution of water recycling systems indicate how this may come about through water treatment advances.

[266] Human Ecology and Conservation

no treatment hopefully are passing. In fact, this old practice could ultimately lead to water recycling treatments. Once environmental water is so contaminated with sewage that it makes intensive purification treatment—in essence the same as sewage treatment—necessary, recycling is inevitable (Fig. 20-4, Option I). If, however, the waste-water treatment is intensified, so that the purity of the water produced approaches the quality of environmental water or of purified water, water recycling will be achieved without the total pollution of the environment (Fig. 20-4, Option II). If the earth's population continues to explode, water recycling may simply be forced upon society by the necessity for water.

The Future

Contaminants get into the environmental waters from all aspects of human society. Home refuse from toilets, washing machines, garbage disposals, and so on makes a large contribution to water pollution. Industries also dump a large variety of wastes into the waters. Stockyard drainage adds animal wastes of all sorts. Farm fields contribute silt, pesticides, herbicides, and fertilizer leachings to the water. Air pollutants get washed from the air into the water at its surface. All sources of contamination must be attacked in order to achieve any widespread halt to water pollution. The balance of this chapter discusses the kinds of tactics that might be necessary in this effort.

Any water clean-up program is going to cost money. Current estimates set the cost of a 5-year water clean-up program at 50 billion dollars. The price for keeping municipal wastes, industrial wastes, and other contaminants from the earth's waters must be paid. The people will have to pay this cost and will only do so when a clean environment means enough to them.

Municipal Waste

Much municipal sewage is still inadequately treated before being discharged into the environment. About 7–10% of the United States municipal sewage does not even receive primary treatment, and about 25% receives only primary treatment. Unfortunately, the secondary facilities are overloaded so that not all of the secondary treatment on the remaining 2/3 of the sewage is effective. Since the technology for sewage treatment is available, it is indeed a disgrace that so much improperly treated municipal waste is still admitted into the earth's waters.

Industrial Waste

The rate of industrial waste treatment must be accelerated. Thankfully, industry has recently been attacking the problem harder, and their current pollution control measures cost billions of dollars annually. The problem will not be solved, however, until **all** industry sets strict contamination standards and figures the cost of the waste clean up as part of their production cost. Complete industrial clean up will never occur as long as a polluting industry can undersell a nonpolluting competitor.

Pesticides

The use of pesticides which persist in the environment must be minimized on a **worldwide** scale. Insecticides such as pyrethrin, allethrin, and rotenone have low toxicities to mammals and short residual lives. Malathion, methoxochlor, and parathion also have relatively short lifetimes. These insecticides should be substituted for the persistent pesticides wherever possible. These pesticides will naturally be less effective than the longer-lasting ones. Their use, however, will have far less drastic effects on the environment and its biological community.

References and Recommended Readings

A Primer on Waste Water Treatment. Document of the U.S. Department of the Interior, Federal Water Pollution Control Administration, 1969.

ALLEN, W. A., and J. W. LEONARD. *Conserving Natural Resources,* McGraw-Hill, New York, 1966.

BENARDE, MELVIN A. *Our Precarious Habitat,* W. W. Norton, New York, 1970.

BORGSTROM, GEORG. *The Hungry Planet,* Collier-Mcmillan, London, 1967.

BORGSTROM, GEORG. *Too Many,* Collier-Mcmillan, London, 1969.

"By-Products Await Desalter." *Chem. Eng. News,* Ind. & Bus. Sect., July 13, 1970 (pp. 18–20).

Cleaning Our Environment: The Chemical Basis for Action. A report by the Subcommittee on Environmental Improvement, Committee on Chemistry and Public Affairs of the American Chemical Society, 1969.

DASMAN, R. F. *Environmental Conservation,* John Wiley & Sons, New York, 1968.

EDWARDS, C. A. "Insecticide Residues in Soils," *Residue Rev.*, **13**, 83–132 (1966).

EHRENFELD, D. W. *Biological Conservation*, Holt, Rinehart and Winston, New York, 1970.

EHRLICH, P. R. and A. H. EHRLICH. *Population, Resources, Environment*, W. H. Freeman, San Francisco, Calif., 1970.

Final Report on Industrial By-Product Recovery by Desalination Techniques. Prepared by Aqua-Chem Incorporated, Waukesha, Wisconsin, for the Department of the Interior, Office of Saline Water, 1970.

GRANT, NEVILLE. "The Legacy of the Mad Hatter," *Environment*, **11**(4), 18–23, 43–44 (1969).

LEAR, JOHN. "The Crisis in Water: What Brought It On?" *Saturday Rev.*, October 23, 1965 (pp. 24–28, 78–80).

LÖFROTH, GÖRAN, and M. E. DUFFY. "Birds Give Us Warning," *Environment*, **11**(4), 10–17 (1969).

MARX, WESLEY. *The Frail Ocean*, Ballantine, New York, 1967.

NASH, R. G. and E. A. WOOLSON. "Persistance of Chlorinated Hydrocarbon Insecticides in Soils," *Science*, **157**, 924–927 (1957).

NOVICK, SHELDON. "A New Pollution Problem," *Environment*, **11**(4) 3–9 (1969).

"Pollution Price Tag: 71 Billion Dollars." *U.S. News World Rep.*, August 17, 1970 (pp. 38–42).

RIENOW, ROBERT, and L. T. RIENOW. *Moment in the Sun*, Ballantine, New York, 1967.

SMITH, GUY-HOWARD (ed.), *Conservation of Natural Resources*, John Wiley & Sons, New York, 1965.

Summary of Registered Agricultural Pesticide Chemical Uses. 3rd ed. Vol. III. Pesticides Regulation Division, Agricultural Research Service, U.S. Department of Agriculture, May 31, 1969.

TAYLOR, G. T. "The Threat to Life in the Sea," *Saturday Rev.*, August 1, 1970 (pp. 40–42).

21 Man and Energy

Both the human body and human society need a constant input of energy to keep "running" since the energy they have eventually escapes as heat and is no longer "useful" to them. The energy source for the human body is food, and its energy content is measured in terms of calories (cal). The body requires about 2000–3000 cal/day to maintain its metabolic activities. The incalculable number of calories required daily by North American society are obtained from water power, natural gas, coal, oil, and nuclear fuel. Food and fuels, then, are essential to life as it is today.

Food

Food not only provides energy for the body's metabolism; it also provides the nutrients or raw materials necessary for its growth and function. A body which does not receive enough energy (calories) is undernourished; one that does not receive enough of the food constituents which are essential for its proper functioning is malnourished. A person can be malnourished even though he is not undernourished. For example, refined sugar is a source of readily available energy for the body. However, a person can not live indefinitely on only sugar and water. From this diet the body cannot obtain any of the proteins, vitamins, or minerals it requires.

As discussed in Chapter 5, proteins are essential to body function and structure. Man requires an intake of proteins in order to build the proteins of his own body. Certain minerals are necessary to the function of important cellular processes. For example, iron is essential to the oxygen-transport system of the blood, and calcium is required in bone building and blood clotting. Each vitamin also keeps the body from malfunctioning. For example, niacin prevents pellegra, vitamin C prevents scurvy, and thiamine prevents beriberi.

Although vitamins and minerals are essential to the diet, they will not be considered separately in the following analysis of the content and production of food. They are required only in trace amounts, and it is assumed they will be contained in the food produced which has the needed caloric and protein content.

Food Production On Land

Most of the food of the world is produced on the land rather than in the sea. The land can be used to raise cultivated plant crops, or it can be used as grazing area to raise animals. The three main plant crops of the world are rice, wheat, and corn. The total annual yield of these three grains is about 3/4 billion tons. All the other grain crops total somewhat over 1 billion tons annually. Each year about 1/3 billion tons of potatoes and 100 million tons of soybeans are also produced. Compared to this, however, the land only produces 20 million tons of animal protein in a year.

Unfortunately, these seemingly astronomical quantities of food are not enough to feed the 3.5 billion people on the earth. Presently there are as many as 1.5 billion under- or malnourished people on earth. At least 1/2 billion people are chronically hungry or starving. It has been estimated that the world-wide *per capita* food energy requirement is 2350 cal/day. The average caloric content of food available per person is less than this however, if losses, economic factors, and the distribution problems are considered. This means that undernourishment, let alone malnutrition, is presently inevitable in the world. In India the average daily food consumption is only 2000 cal per person. In the United States, on the other hand, each person averages 3000 cal daily.

As long as the world population keeps rising, the goal of improving the world's food supply to the necessary 2350 cal *per capita* level looks dim. Food production *is* increasing, but it is not increasing as fast as the population. In the decade from 1956–1966, the food production *per capita* for Africa and South America declined slightly, but remained about constant for the Far East (excluding Red China) even though the absolute food production increased in all these countries. In India the population increased about 2.5% yearly, but the food production increased by a slightly smaller percentage during the same period.

Man and Energy [271]

It is, of course, misleading to consider only the caloric content of the world food production since nutritionally the protein content is also important. The world nutritional picture is not nearly as optimistic as that painted by caloric considerations alone. The majority of the world's population today have a minimum, or below minimum, protein uptake.

The best sources of protein are meat, milk, eggs, and fish—all animal products. Cereal grains and soybeans are the plant crops richest in protein. These plants can furnish a large share of the protein intake of the human body, but animal sources furnish a better, more concentrated, form of protein. The raising of animals, however, is a much less efficient method of food production than is the growing of plants. Plants are able to synthesize all the food they require for their growth and development from inorganic materials readily available in the environment, but animals are not. The food animals provide represents only about 10–20% of the food they required for their life and growth. Since there is an average of 80–90% food energy loss at every link in a food chain, the most efficient means of animal protein production is the raising of herbivores (Fig. 21-1). (This is, of course, only 10–20% as efficient as the growing of plants.)

A look at the world's land resources does not provide any obvious

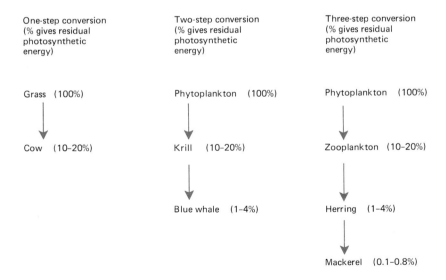

21-1 **Conversion efficiency of plant material to animal protein** (assuming a conversion of 10–20% with each step.) This step sequence results in major reductions of the available photosynthetic energy for each higher feeding level.

[272] Human Ecology and Conservation

solutions to the food shortage. Most of the really productive cropland has already been pressed into the service of food production. Only 1/4 of the earth's 32 billion acres of land surface is considered "potentially" useable for cropland. Most of these 8 billion acres can probably only be useable after intensive work or with techniques not now available. About half of this land is in the tropics. This area has discouraged farming for many reasons. The hot climate and high humidity there accelerate the break down of organic nutrients in the soil and make it less productive. In addition, the soil is often leached by heavy tropical rains and is often highly acidic.

About 8 billion acres of the world's land area which is not adaptable to cultivation because of the soil content, terrain, or rainfall is adaptable to grazing. Land use of this kind would provide a good source of the proteins necessary for the human diet. However, the raising of herbivores is such an inefficient mode of food production that this land use shows little promise of being able to keep an exploding population well fed.

Food Production In The Seas

The oceans are *not* a limitless source of human food. At present only about 60–70 million tons of animal protein are obtained from them per year. This represents only a few per cent of the total caloric intake of the human population. It is estimated that the maximum fish production maintainable from the oceans is between 100–150 million tons annually, or about twice the current production. This maximum figure would satisfy only about 20–30% of the minimum protein requirement for the present growing human population and an even smaller percentage of its caloric needs.

The unexpectedly low maximum fish production estimate for the oceans results from the nature of food production in the ocean. Plants can only grow near the ocean's surface where enough sunlight is available for photosynthesis to occur. The nutrients in the ocean, however, which these plants also require are not generally at the surface. Only about 10% of the oceans are reasonably productive, usually near seacoasts or in areas where upwelling brings nutrients from the ocean's depth to the surface. In addition, food chains for edible ocean animals are longer than those of terrestrial herbivores (Fig. 21-1). Even the relatively direct conversion of phytoplankton to krill to blue whale is a two-step chain. This means only about 1–4% of the plant food is converted to animal protein. A more typical oceanic food chain is that of the mackerel. The conversion of phytoplankton to zooplankton to herring to mackerel is a three-step chain. Based on typical conversion factors, it would take 1000 lb of phytoplankton to produce 1 lb of mackerel.

It seems highly impractical to harvest the plant life of the ocean, the plankton, rather than the animal life in an effort to increase the efficiency of food production of the ocean. The physical problems of harvesting these microscopic organisms from such large volumes of water would be tremendous. In all likelihood more energy would be used in the harvesting process than would be obtained in the product, even in the most productive parts of the ocean. Aside from harvesting problems, a total harvest of all the ocean's plankton once a year would produce less total tonnage than the annual grain harvest from land. Plankton removal on this scale would undoubtedly disrupt the oceanic ecosystem and could upset its current fish production. Besides, foodstuffs made from plankton may not taste very good.

Even the present level of food production of the oceans may be endangered by exploitation and pollution. Either one of these trends could seriously affect the briny ecosystems and decrease their productivity. Currently neither of these dangers are prohibited by international law.

The effects of exploitation are, unfortunately, fast becoming evident in the Atlantic Salmon population. These fish feed in the Atlantic and spawn in the freshwater rivers of North America, Great Britain, and Norway. When their feeding grounds were unknown, the salmon population was regulated since each country regulated the kill of the spawning salmon in their own coastal waters. In 1960 the feeding ground was discovered off the coast of Greenland within the territorial waters of Denmark. From 1963–1968, 2 million salmon were taken in this feeding ground. The spawning runs in North America and Britain declined. Diplomatic attempts to at least limit the take off Greenland were rebuffed by Denmark. In 1965 the Danes started open sea fishing as well. Their annual open sea catches increased, and in 1969 they caught 628,000 fish—4.4 million lb. It is now estimated that if this rate of catch for spawners is continued for many more years, the species will be extinct. Both Denmark and West Germany did not accept the recommendations made by the Northeast Atlantic Fisheries Commission and the International Commission for Northwest Atlantic Fisheries to ban salmon fishing on the open seas, however.

This example of species exploitation illustrates the problems that prohibit the efficient management of the oceans. The territorial waters of a country can be "used" precisely as that country decides. The open seas can be used as any sovereign country decides. At present there is no form of control that insures ocean management in ecological interests.

Water pollution is also a real threat to the food production potential of the oceans. As already mentioned, the most productive ocean areas are the coastal waters. The productivity of these coastal areas depends in part on the nutrients brought there from inland rivers. These rivers are now bringing contaminants as well as nutrients to the coastal waters. It may seem that the oceans are too big to pollute, but phenomena like biological concentration indicate that indeed they are not. It has been found that DDT, present in concentrations far below 1 ppm, can reduce photosynthesis in some algae. Unless the quality of the ocean is preserved it may be impossible to maintain its current level of food production, let alone reach more optimistic levels.

Synthesis of Food

The synthetic production of nutrients will undoubtedly increase and may someday provide a significant percentage of the world's food. Using methods similar to those now available, however, the synthesis of foodstuffs cannot solve the current food-supply problems nor meet the needs of the increasing population alone. The problems and requirements of producing food synthetically on so large a scale are enormous. The total volume of food needed for the world population is far beyond the present, or foreseeable, capabilities of the chemical industry. The total annual synthetic organic chemical output of the United States chemical industry is measured in tens of millions of tons. The total annual food consumption, on the other hand, is measured in billions of tons. Even if the chemical industry had the capacity, it would soon run out of raw materials. Unless a method is developed which mimics photosynthesis, food synthesis would require organic "starting" materials such as coal and crude oil. Synthesis on this scale would place a huge drain on these nonrenewable resources and would by no means represent a long-term solution to the food problem. Chemical processes of this sort also require energy. A major synthesis of foods, then, would tax considerably the energy reserves now available. Water would undoubtedly be needed in this process, and water supplies are even presently under strain.

All these considerations indicate that the synthesis of foodstuff by methods presently available will undoubtedly be limited to special products, at least for the near future. It is not currently considered an effective means for producing the world food supply. If a totally different kind of synthesis were developed which, like photosynthesis, was fueled by solar energy and utilized inorganic materials, CO_2, and H_2O, large-scale food synthesis might become a reality. At the present state of technology, however, this eventuality seems unlikely.

Microbiological Production of Food

Some microorganisms can live on inorganic compounds and some can live on waste organic materials. For example, organisms of high protein content, yeast, can grow on paper mill wastes. Microbial conversions of this kind could be a way to produce protein foodstuffs from the always readily available organic waste products. Microbiological food production of this sort looks promising for the future. Ways must be developed, however, to separate the food materials produced by the microorganisms from the nutrient waste medium so that the food materials would be acceptable to the populace.

Fuels

Our technological society requires energy in cheap and abundant supply. The primary sources of this energy today are coal, petroleum fuels, and nuclear fuels. Although water power is used widely for the generation of electricity, it plays a relatively minor role in the overall pattern of world energy production. The dependency of "civilization" on the organic and nuclear fuels makes their availability and use an important aspect of the ecology of social man.

Coal

Coal is a fossilized form of plant material. The remains of plants which grew millions of years ago, deposited in thick layers underground, were subjected to intense heat and pressure deep within the earth. This process, occurring in the absence of O_2, formed coal. Coal is, therefore, a very concentrated form of organic compounds which were photosynthesized by plants millions of years ago. The time required for this transformation makes coal a nonrenewable resource.

The uses of coal are many. As an energy source it provides heat, generates electricity, and powers industry. In addition it is used as a raw material in the production of such chemicals as methane, hydrogen, naphthalene, benzene, and xylene. Steel is also produced with the use of coal in the form of coke.

The coal supplies are estimated to be quite large. The world's coal reserves contain enough coal to last 300–400 years if other energy sources are utilized as well. If coal were to be the sole energy source, however, these reserves would last only another 100–200 years.

Coal, unfortunately, is not as clean a fuel as petroleum. Because of this, petroleum fuels are being used more and more by industry. The technology of coal utilization must be improved in order to utilize fully these vast coal supplies without further polluting the environment. For instance, a method to remove the sulfur from either the coal itself or the smoke it produces upon combustion must be developed if sulfur and its oxides are to be kept out of the atmosphere.

Petroleum Fuels

Like coal, the petroleum fuels are products of plants which grew millions of years ago. These ancient plant deposits were subjected to heat and pressure in water without much oxygen. As a result of this process a vast mixture of hydrocarbons was formed. This mixture contains compounds with various numbers of carbon atoms in their molecular structures—for example, methane (CH_4), ethane (C_2H_6), propane (C_3H_8), butane (C_4H_{10}), and so forth. The compounds with few carbon atoms in their molecular structure are gases and comprise what is known as natural gas. The petroleum hydrocarbons whose molecular structures are composed of longer carbon skeletons are liquids. This mixture of liquid hydrocarbons is known as crude oil. The time required for the production of these compounds makes petroleum also a nonrenewable resource.

Petroleum fuels supply approximately 2/3 of the energy that runs the North American industrial complex. Crude oil is the raw material for the production of the gasoline, jet fuels, and the fuel oils that power the world's transportation systems. It is also used as starting material for the synthesis of nylon, rubber, and most plastics. Refined crude oil yields lubricating oils, greases, vasoline, tar, and other everyday products. Natural gases are used in heating and cooking. If this current heavy use of petroleum fuels continues, the known petroleum reserves will dwindle about the year 2000.

Heavy oil sands and oil shales may provide an alternative source of organic fuels and chemical raw materials. Heavy oil sands contain large amounts of petroleumlike materials which cannot be recovered as fluid without being refined from the parent materials. Oil shales contain organic materials which differ chemically from crude oil but can be refined into useful products. The estimated reserves of heavy oil sands are large. At present the amount of organic products recovered from these two sources is limited due to the technology necessary for this recovery. Much of the oil shale is located in Colorado, Utah, and Wyoming. In these areas the limited water supply hampers the development of the necessary refineries. Probably neither the sands nor shales will ever furnish the quantities of petroleum needed for an industrial energy source. They may become, however, an important source of organic materials for chemical synthesis. If this occurred, the petroleum fuel could then be used primarily for energy.

Nuclear Fuels Nuclear energy is the only energy source (of those currently capturable) capable of providing the amount of energy needed to maintain a technological society in the future. At the current rate of consumption of coal and oil the fossil fuel reserves will be depleted well within 400 years. During this time nuclear energy sources must be developed and perfected if the industrial society is to continue to function without interruption. The current production of nuclear energy cannot as yet supply the quantity of energy needed by a technological society.

The type of nuclear reaction in use today is based merely on the "burning" of the nuclear fuel ^{235}U. This uranium isotope undergoes fission to produce radioactive wastes, neutrons and a large amount of energy:

$$^{1}n + {}^{235}U \xrightarrow{\text{fission}} \text{several } {}^{1}n + \text{radioactive wastes} + \text{energy}$$

During fission, 1 gram (g) of ^{235}U (1/454 lb) produces the same amount of heat energy as would the burning of 3 tons of coal or 14 barrels of crude oil. Since neutron bombardment starts the ^{235}U fission reaction, and neutrons are produced in the fission reaction, the reaction is self-perpetuating. It will continue until all the fuel is consumed. Although the technology to use this process is available, the supply of ^{235}U is small. The estimated ^{235}U supply would be used up in less than 25 years at the current rate of growth of nuclear energy production.

It is hoped, however, that a "breeder" reactor can be perfected which will be able to supply an unlimited amount of nuclear fuel. In this reactor the neutrons produced by the fission of ^{235}U can "breed" ^{238}U or ^{232}Th (thorium-232) into the ^{239}Pu (plutonium-239) and ^{233}U nuclei, respectively. The ^{238}U and ^{232}Th nuclei are not fissionable, but the ^{239}Pu and ^{233}U they breed are readily fissionable:

Breeding reactions (start with $^{235}U \xrightarrow{\text{fission}} {}^1n$):

$$^1n + {}^{238}U \longrightarrow {}^{239}Pu$$
$$^1n + {}^{232}Th \longrightarrow {}^{233}U$$

Fission reactions:

$$^1n + {}^{239}Pu \longrightarrow \text{neutrons} + \text{wastes} + \text{energy}$$
$$^1n + {}^{233}U \longrightarrow \text{neutrons} + \text{wastes} + \text{energy}$$

The breeding reactions, then, produce fuels that are more easily fissionable than the starting materials, ^{238}U and ^{232}Th.

Breeder reactors could produce an "unlimited" supply of energy because the materials they render useful as nuclear fuels are in rich supply. Uranium ore contains less than 1% of the fissionable ^{235}U isotope and over 99% of the unfissionable ^{238}U. A breeder reactor therefore makes over 100 times as much uranium fuel available. In addition it turns ^{232}Th into a nuclear fuel. Thorium compounds are not useable at all in a burner-type reactor. One deposit of granite in New Hampshire contains enough thorium to provide energy equivalent to 20 times the total original United States coal reserves and 750 times the original United States oil deposits.

The Future

The world population is increasing. It is uncertain whether the production of the energy necessary to keep this population alive and its social structure functioning can increase at an equal rate, let alone faster. Future population sizes ultimately depend on the energy production, so it is impossible to predict the nature of future populations. It is interesting, however, to speculate on possible future population trends which could result from different energy-production capabilities. These population trends are readily compared by a graphic representation of population size versus time (Fig. 21-2). The possible trends can be called the "technological curve," the "animal curve," the "controlled curve," and the "disaster curve."

If technology can find ways to produce unlimited nuclear energy, an unlimited nutritional food supply, and ways to clothe and house unlimited numbers of people, the population can continue to grow as represented by the "technological curve." These are big "ifs." Even if they are all realized without causing insurmountable prob-

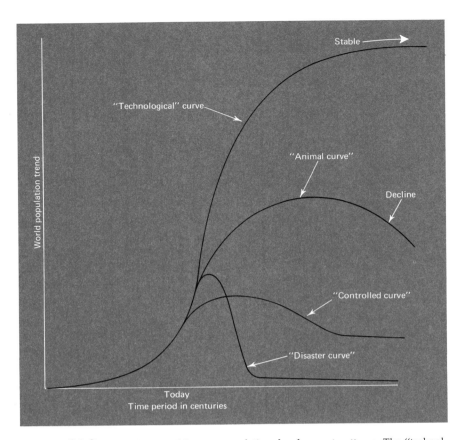

21-2 **Speculations of human population development patterns.** The "technology" curve assumes technological miracles that eventually reach the carrying capacity of the earth. The "animal" curve assumes the maximum population the earth can carry short of "technological miracles." The "optimum" curve assumes a rational limitation of population at something less than maximum survival numbers. The "disaster" curve assumes a collapse of the environment and probably of the social structure.

lems such as the safe disposal of enormous amounts of radioactive, organic, and industrial wastes, there will ultimately be a point where the population must be controlled so as not to exceed the earth's carrying capacity. The "technological curve" will then flatten out as the population stabilizes. Even though the human reproductive capacity is unlimited, the earth's capacity to support a human population is not.

The population of a species which cannot "control" its environment is shaped by the energy available to it and its natural predators. Since man has few "natural" enemies left uncontrolled, his population will increase until it cannot obtain sufficient energy. If technology cannot produce unlimited food and nuclear energy, the human population will naturally increase to the point of mass starvation. This possibility is represented in the "animal curve." Once this point is reached the food energy sources become overused and misused. The earth's carrying capacity will decrease. Starvation will continue until the population is within this lower level. The degree to which the waters are polluted, the air contaminated, the oceans exploited, the forests overcut, and the land eroded indicates that man decreases the carrying capacity of the earth just as other animals do.

Man can, however, set his own population controls so that his population is limited by birth control methods rather than by energy supply or the earth's carrying capacity. If he does this, his population will stabilize, decrease, or slowly increase depending on whether he commits himself to zero, negative, or controlled population growth, respectively. One such trend is depicted with the "controlled curve." Binding birth control in any form has never been "popular," however. This trend may, indeed, be impossible.

A sudden or violent collapse of the carrying capacity of the environment will force the human population trend exhibited in the "disaster curve." There are many potential causes of such a collapse—for example large scale nuclear, chemical, or biological warfare; the pollution and ultimate death of the oceanic ecosystem; or the gradual accumulation of radioactive materials in the environment.

Of these population trends, the "controlled" and the "technological" curves seem the most desirable since they do not involve loss of human lives. In deciding which course to take, although both paths may be difficult to implement, man should consider the ecological changes that an even larger human population will bring about. Surely more trees will be hewn to increase cropland and provide more building materials. Scenic areas will be terraced and leveled for housing and farmland. Cities will expand skyward or underground in order to house more people without taking land necessary for food production. Wild animals will be replaced with cattle, sheep, and goats as these animals are better sources of food. Eventually the vertebrate population will be reduced to man and his domesticated animals. Many other such changes will be forced on the earth's biosphere, and man must decide whether an increased population is worth it. Hopefully he will make his decision before either the "animal curve" or the "disaster curve" are forced upon him.

References and Recommended Readings

ALLEN, W. A., and J. W. LEONARD. *Conserving Natural Resources*, McGraw-Hill Book Company, New York, 1966.

BORGSTROM, GEORG. *The Hungry Planet*, Collier-Macmillan, London, 1967.

BORGSTROM, GEORG. *Too Many*, Collier-Macmillan, London, 1969.

BROOKS, R. R. "People Versus Food," *Saturday Rev.*, September 5, 1970 (pp. 10-14, 33).

EHRLICH, P. R. and A. H. EHRLICH. *Population, Resources, Environment*, W. H. Freeman, San Francisco, Calif., 1970.

EHRLICH, P. R. and A. H. EHRLICH. "The Food-From-The-Sea Myth," *Saturday Rev.*, April 4, 1970 (pp. 53-65).

GRAHAME, ARTHUR. "Last Chance for Atlantic Salmon?" *Outdoor Life*, June, 1970 (pp. 41-43, 92, 94, 96, 98, 100).

Resources and Man. Committee on Resources and Man, National Academy of Sciences, National Research Council, W. H. Freeman, San Francisco, Calif., 1969.

RYTHER, J. H. "Photosynthesis and Fish Production in the Sea," *Science*, **166**, 72-76 (1969).

TAYLOR, G. R. "The Threat to Life in the Sea," *Saturday Rev.*, August 1, 1970 (pp. 40-42).

WEINBERG, A. M. and R. P. HAMMOND. "Limits to the Use of Energy," *Amer. Sci.* **58**, 412-418 (1970).

WURSTER, C. F. "DDT Reduces Photosynthesis by Marine Phytoplankton," *Science*, **159**, 1474-1475 (1968).

Man and Shelter 22

Nearly all warm-blooded animals get protection from the elements with both body cover and physical shelter. Their bodies are covered with fur or feathers and they obtain shelter in all kinds of places—in holes in the ground or trees, thickets, brush piles, grass nests, or windward ravines. The biological requirements of modern man for body cover and shelter are now integrally intertwined with social traditions, technological advances, and fashion. Nearly every environmental resource is touched in the fulfillment of man's need for cover.

Clothing

Man's clothing is made from plant and animal products as well as from synthetic chemical products. The production of all these materials makes demands on the environment.

Plant Products

Cotton and linen are the oldest forms of cloth made from plants. Cotton is made from the cotton plant and linen from the flax plant. Rayon, a cloth developed in modern times, is a derivative of wood. The manufacture of all these fibers, requires land to grow the necessary cotton, flax, and trees. These products are considered "renewable" since their sources can be continually replanted. Nowadays not much flax is raised in North America. Cotton, however is a major crop in the United States. In 1968 the amount of land devoted to cotton was 10 million acres; 5 billion lb of cotton were produced. The cotton crop, therefore, involves extensive land use and often requires a heavy application of insecticides. To make rayon the cellulose of trees is dissolved chemically. When this cellulose solution is forced through small openings into a H_2SO_4 (sulfuric acid) bath, strands of cellulose are formed. These strands are spun and woven into rayon fabric. The production of rayon, requires that land be retained as forest. If these forests are properly managed, rayon production will not harm the environment since forests do not damage the soil.

Animal Hair Products

The wool from sheep is the primary animal hair product used today in clothing. In 1967 the shearings from 22 million sheep yielded 185 million lb of wool. During that year the United States *per capita* wool use was about 2 lb. Angora wool from Angora rabbits and mohair from goats are also used in clothing, but in much smaller amounts. All the hair now used in clothing has been removed from the animal's skin. This hair is then formed into thread and woven into cloth.

Since sheep grow new wool each year, wool is also a "renewable" product. Its production is not dangerous to the environment if the sheep's grazeland is well managed. Too many sheep enclosed in one area can destroy their range by eating all its vegetation. An overeaten range cannot support any wildlife and is subject to erosion.

Leather Products

Leather has been used for clothing for a long time. Early North American societies used the skins of buffalo, moose, elk, deer, seals, and other animals to clothe themselves. Because it is tough, workable, pliable, and pleasing to the eye and touch, leather is still used today in shoes, boots, belts, handbags, and jackets.

Tanned cowhide is the main leather in use today. It is obtained primarily from the more than 100 million cattle in the United States. Although the cattle have to be killed to obtain their hide, their flesh is also used as an important source of animal protein. The use of cowhide is not damaging to the environment if the cattle are not allowed to overgraze their ranges.

Synthetic Fibers

Synthetic fibers such as nylon, dacron, and orlon, are now commonly used in clothing. These fibers, polymers produced by chemical processes, can be formed into threads and woven into cloth. Nylon is made from organic materials derived from petroleum.

Chemically it is a copolymer, as it is a polymer made up of two different types of monomers. One monomer is an organic acid; the other is an organic base. These monomers are joined together by amide linkages:

$$n \; \mathrm{HOOC\text{-}CH_2\text{-}CH_2\text{-}CH_2\text{-}CH_2\text{-}COOH} + n \; \mathrm{HN(H)\text{-}CH_2\text{-}CH_2\text{-}CH_2\text{-}CH_2\text{-}CH_2\text{-}CH_2\text{-}NH} \xrightarrow{\text{polymerization}}$$

$$\left(\mathrm{\overset{O}{\overset{\|}{C}}\text{-}CH_2\text{-}CH_2\text{-}CH_2\text{-}CH_2\text{-}\overset{O}{\overset{\|}{C}}\text{-}N(H)\text{-}CH_2\text{-}CH_2\text{-}CH_2\text{-}CH_2\text{-}CH_2\text{-}CH_2\text{-}N(H)} \right)_n$$

Dacron is a polymer called a polyester since it is a copolymer of organic acids and organic alcohols held together by ester linkages:

$$n \; \mathrm{HOOC\text{-}C_6H_4\text{-}COOH} + \mathrm{H\text{-}CH_2\text{-}CH_2\text{-}H} \; (\text{with } \mathrm{HO, OH}) \xrightarrow{\text{polymerization}}$$

$$\left(\mathrm{\overset{O}{\overset{\|}{C}}\text{-}C_6H_4\text{-}\overset{O}{\overset{\|}{C}}\text{-}O\text{-}CH_2\text{-}CH_2\text{-}O} \right)_n$$

Orlon is a polymer made up of just one kind of monomer.

$$n \; \mathrm{CH_2{=}CH\text{-}CN} \xrightarrow{\text{polymerization}} \left(\mathrm{CH_2\text{-}CH(CN)} \right)_n$$

The clothing made from these synthetic fibers has advantageous properties. For example, nylon is strong, yet very light; dacron is crease resistant; and orlon is unchanged by sunlight. The production of these fibers, however, taps the natural sources of organic raw materials, particularly petroleum. The synthetic textile industry also requires much machinery, manpower, and energy to function. Hence these fabrics place demands on the earth's reserves of fossil fuels and organic raw materials rather than on the soil and vegetation.

Man and Shelter

Housing

Early man found shelter in caves where he built a fire for warmth. As time went on men began to erect their own shelters using the natural materials in their environment. They used such materials as snow, buffalo hide, grass, adobe bricks, sod, logs, and stone. Man gradually learned more and more about structure and the capabilities of the natural materials. He then began to reshape and mix these materials so they would be useable for specific structural purposes. Today architectural traditions and construction know-how have both developed to a point that the natural resources of not only North America but *other countries* as well are utilized in the construction of even a modest structure in the United States. An analysis of construction materials used in such a shelter will indicate the complex interactions between man and his environment brought about merely by man's need for cover.

Concrete Many structures have a concrete foundation. Concrete is a mixture of sand, gravel, and cement. The cement, made from mixing calcium compounds and clay, contains calcium silicates, calcium aluminates, silica, and aluminum silicates. This mixture will form a rock hard mass once it is mixed with water and allowed to set. Since it can be reinforced by iron rods and molded into a desired shape, concrete forms a very rigid structure of any design when it dries. These properties lead to its use in house foundations and the skeletons of larger buildings. Over 70 million tons of cement are produced annually. The raw materials for concrete—limestone, clay, sand, and gravel—are still in relatively abundant supply.

Lumber The framework of most houses is made with lumber; the floor joists are commonly 2 × 8 or 2 × 10 in. planks, the wall studs, 2 × 4, and the rafters, 2 × 6 or 2 × 8. In addition, plywood is usually used to cover the roof and floor and for porches and eaves. The lumber used in housing is usually from pine, fir, spruce, Douglas fir, or hemlock trees. The estimated total lumber consumption for 1968 was about 41 billion board-ft. Of this 36.5 billion board-ft were produced in the United States alone. Another 20 billion sq ft of plywood were used. The *per capita* consumption is 205 board-ft in lumber and 40 board-ft in plywood. These extensive demands on the forests for wood products, for housing as well as other purposes, make the wise management of forests essential.

Fiber and Composition Board

Fiber and composition boards are made from ground and chopped wood, respectively. These wood scraps are mixed with a binding agent and formed into sheets. Nowadays houses are commonly enclosed with fiber board and floors are often covered with composition board. The siding on many modern houses is often made of these materials as well. The increased use of these lumber substitutes which are made of scrap wood indicates an efficient use of forest timber. These "boards" are still wood products, however, and their continued production also demands good forest management.

Flooring

The wood flooring laid over a plywood subfloor is a high quality hardwood such as oak, birch, or maple. It is usually clear of knots and other defects so it is produced only from high quality trees. Forests must indeed be carefully managed to insure the growth of such quality trees.

Plasterboard

Sheets of plasterboard are commonly used to cover the interior walls of a house. Plasterboard is a sheet of plaster covered on both sides with paper or cardboard. Plaster is composed of calcium oxide, a limestone product that is much softer than cement. Neither the limestone or the paper are in short supply, but the paper production, once again, depends on good forest management.

Glass

Glass is used in a house for windows, mirrors, and insulation. It is a product of silicon dioxide (sand) in combination with compounds of calcium, sodium, oxygen, aluminum, and iron. Its transparency in sheet form leads to its use in windows and mirrors. When spun, however, it provides excellent insulation. Foil-wrapped layers of spun glass are often placed between walls and in ovens and refrigerators for this purpose. The raw materials necessary for its production are in large supply in the environment. Millions of tons of selected sand are used in glass production annually.

Steel

Because steel is very strong and durable, it has been put to many structural uses in housing. Many structures have steel supporting beams. Concrete building foundations are generally reinforced with steel rods. The nails and staples which hold house parts together are also steel.

Steel is a mixture of iron and carbon. Its properties are determined by the amount of carbon and other metals alloyed with the iron. It is made by heating iron ore, an impure iron compound, usually an oxide, with coke and limestone. This treatment forms two molten liquids which are separable—molten iron and a molten mixture of impurities called slag.

Iron is produced from iron ore at a rate of 90 million tons per year in the U.S. To make steel some scrap iron is also used. The total yearly steel production is approximately 130 million tons. The iron ore reserves in this country would probably be depleted in 50 years if these were the only sources for this ore for the United States. The importation of iron ore from Canada, Venezuela, and other countries will extend the lifetime of these reserves.

Steel alloyed with chromium and nickel resists rust and corrosion and became known as stainless steel. It is commonly used in houses for sinks and other fixtures which are frequently in contact with water or food. The United States is almost completely dependent on foreign supplies for both nickel and chromium.

Steel that is coated with zinc, galvanized steel, is resistant to rust by water. Galvanized steel is used in eave troughs because of this resistance. Zinc demands in the United States cannot be met by domestic zinc reserves. In 1968 the United States imported approximately half of the zinc it consumed.

Aluminum Since aluminum does not rust or corrode, it is used for window frames and outer doors. The domestic aluminum resources are not large, and the United States currently imports this metal to help it meet domestic demands. In 1968, 1/3 of the aluminum consumption was supplied by imports or scrap recovery. The dependence on foreign aluminum sources for both present and future needs makes the recycling of aluminum desirable whenever possible.

Copper The electric wiring and water pipes in houses are generally made of copper. United States copper reserves cannot meet domestic demands. This country imports a great deal of copper from Chili, Canada, Phillipines, Peru, and Africa and recovers copper from scrap to meet its needs for this metal.

Clay Aside from its use in cement manufacture, clay is also used in the production of brick and tile. The brick used for outside construction, inside walls, and fireplaces is made from relatively impure clay. The hardness and glaze of bricks are determined by the degree to which the clay is fired. The porcelain and fine tiles used for bathroom walls, floors, and fixtures are made from a higher grade of clay. The firing of this clay is controlled so as to render these materials hard and highly glazed. Clays are relatively abundant, but current processing practices can cause pollution of the environment.

Solder is a metal alloy with a low melting point that is used to make electrical connections and to seal pipe connections. It is usually composed of tin and lead. This country is entirely dependent on other countries for tin. The primary sources for this metal are Malaysia, Thailand, and Bolivia. Lead is produced in the United States and is imported chiefly from Canada, Australia, and Peru.

Many products used in housing—tar, asphalt, plastics and rubber—are made from petroleum. Tar is commonly used to coat concrete foundations. It protects the concrete from constant exposure to the water in the soil. Asphalt provides the "tar paper" foundation for shingles and linoleum. Some floor tiles and counter tops are made of plastics. Even synthetic rubber is used in housing. Latex rubber based paints are receiving widespread use, both indoors and outdoors. All these products must be chemically processed from petroleum. Their use in housing puts an additional drain on the petroleum and other fuel reserves.

Petroleum Products

A single house, then, is clearly composed of materials from forests and other natural resources which are processed by all kinds of industry—chemical factories, refineries, smelters, blast furnaces, and so on. These materials come from the United States and the *rest of the world* as well. Such far reaching environmental interaction in satisfying an organism's need for cover is unique with industrialized man. As the population grows, its need for shelter will also increase. Fulfillment of this need by current environmentally demanding practices may put a dangerous strain on the earth's resources and carrying capacity.

The Future

The ways in which man fulfills his need for shelter merely *exemplifies* how extensively industrial man interacts with the materials other than air and water in his environment. Man's need for shelter does not *require* this extensive interaction, but his industrial society has made it *possible*. Now man takes for granted materials which are only available through import or technology. As the population grows, however, this extensive interaction with the materials of the environment will only be increased. In time, the material resources of the earth may be seriously depleted and the earth a veritable garbage dump of used products. This situation can only be combated by proper management of the earth's "renewable" resources and the recovery wherever possible of the "unrenewable" ones.

Management of "Renewable" Resources

The renewable resources, such as forests, livestock, plant crops, grazing range, and so forth, will be continually available if they are carefully managed. The needs of the organisms involved must be carefully considered from an ecological viewpoint in this management. The forest that is managed for close utilization of various sized wood products and for good reproduction at the maturity of the stand will produce continually. Similarly, the range that is grazed at, or below, its carrying capacity will continue to support the animals necessary for food, wool, and leather.

Metal Recycling

The earth's metal resources are limited. The elemental metals used in many products are generally stable and do not decompose. Unless these "used" metals are dispersed too widely throughout the environment, their recovery should be possible.

At present the recovery of metals is only undertaken if it is economically feasible. The only metals currently being recycled are those which can be recovered at a *profit,* regardless of their known reserves. Many metals are still only used once in spite of the dwindling reserves. Economics will eventually force the recycling of most metals. As the metal reserves get smaller, the demand for these metals and the cost to mine them will raise their price. At some point the price of metals will be high enough to make their recovery from used products economically feasible. This economic force may not be applied in time, however, and the metals may be too widely dispersed throughout the earth ever to be recovered.

Man's industrialization has indeed created "havoc" with his ecosystem. As industry grew and tapped more and more resources, the population could continue to grow. Technology supplied this population with all manner of conveniences and services which have come to be a integral part of life. The strain this puts on the environment, however, is becoming obvious. Naturally, it is desirable to try to ease the strain on the environment without giving up too many of the goods and practices to which society has become accustomed. This may be possible to some extent if some corrective measures are made in time. Some of these measures will no doubt be "inconvenient." The "inconveniences" of saving, separating, and transporting tin cans and other metal products, paper, bottles, plastic goods, and other materials for recycling, however, may seem small indeed compared to the "inconveniences" that would be suffered by *all* organisms of a totally disrupted and polluted biosphere.

References and Recommended Readings

Agricultural Statistics 1969. U.S. Department of Agriculture. U.S. Printing Government Printing Office, Washington, 1969.

ALLEN, W. A. and J. W. LEONARD. *Conserving Natural Resources,* McGraw-Hill, New York, 1966.

CRAM, D. J. and G. S. HAMMOND. *Organic Chemistry,* McGraw-Hill, New York, 1964.

Minerals Yearbook. U.S. Department of the Interior, U.S. Bureau of Mines, 1968.

SORUM, C. H. *Fundamentals of General Chemistry,* Prentice-Hall, Englewood Cliffs, N.J., 1955.

The Demand and Price Situation for Forest Products 1968–1969. U.S. Department of Agriculture, Forest Service, Miscellaneous Publication no. 1086, 1969.

Timber Trends in the United States. U.S. Department of Agriculture, Forest Service, Research Report no. 17, 1965.

Part VII

On Life, Biology, and Science

On Life, Biology, and Science 23

A question basic to biology, the study of living things, is, "What Is Life?" This question has been put off until now since some of the concepts presented in the preceding chapters help in attempting to answer it. There is such a wide diversity amongst all the structures on this earth that there is no sharp line between living and nonliving objects. The delineation does not seem so fuzzy in the macroscopic world, but many microscopic systems exist which are hard to classify as either living or nonliving. This situation is almost to be expected if living things evolved from nonliving things. It is, therefore, impossible to give an absolute definition to "life." In spite of this quandry, it is useful to develop a working definition of "life." One approach is to look at the forms that are intuitively felt to be living and determine the characteristics which they have which set them apart from those forms felt to be nonliving.

What Is Life?

The most general characteristics of "living" things are summarized below:

1. Living systems are composed of many different chemicals which interact with one another in a complex system of chemical reactions. This collection of chemical reactions which occurs in a living system is referred to as metabolism. Metabolic reactions only continue to occur if energy is supplied to the system. The sum of, and expression of, these reactions is the form and function of

Earth rise on the lunar horizon. (*Courtesy of NASA, Houston, Texas.*)

the living organism. In general, metabolic reactions depend upon the availability of certain enzymes which are synthesized within the organism based on the information in the nucleic acids. The total chemical structure and function of a living system, therefore, is dependent on chemically coded information in the form of nucleic acids.

2. Living systems usually obtain the energy used in metabolism from the chemical break down of organic food materials formed ultimately by photosynthesis. During this break down, the energy released is concentrated and stored in compact and high-energy chemical structures such as ATP.

3. Living systems grow by synthesizing the necessary components from the chemical raw materials they obtain from their environment in energy-consuming chemical reactions. The compounds synthesized by each organism as it grows are characteristic for that organism and its species.

4. Living systems move by the utilization of some of the chemical energy they obtain during the break down of energy-rich organic compounds. This chemical energy is utilized within the organisms. The movements of nonliving systems all result from gravitational or mechanical forces applied outside the system.

5. Living things fall into general groupings called species. Speciation arises from the variations between organisms resulting from slight chemical differences between them and natural selection caused by the way these organisms interact with their total environment.

6. Living systems generally respond in specific ways to environmental components. For instance, the leaves of a plant face toward light. Its stem opposes gravity, and its roots follow gravity.

7. Living things can reproduce their species. Sometimes not all the organisms of a species can reproduce, but the species is maintained because at least some individuals within it do have the ability to reproduce.

Some objects are easy to classify as either living or nonliving using these characteristics as criteria for "life." For instance, using these criteria, a man is most assuredly alive, but a rock is not. Man is indeed a complex system composed of interacting chemicals which requires an energy source for his maintenance. He gets this energy during respiration, the aerobic break down of the food he eats. The rock, however, has no energy requirements and performs no chemical functions. Man grows by energy-requiring synthetic reactions; the rock does not grow except by crystallization. A man's movement results from energy released during chemical break down within his cells; the movement of a rock results from external forces. Man is a member of a species which originated from organismal variation and natural selection; the rock is not. Man responds to many environmental factors; the rock responds to few. Man reproduces a likeness of himself; a rock cannot.

Is Virus Living?

These criteria, when applied to either a rock or a man seem quite helpful in their classification as to whether an object is "alive" or not (Table 23-1). Other forms are not as easy to characterize as living or nonliving, however. The virus is one such borderline case. The "life cycle" of the bacteriophage T_5 virus has been extensively studied. The criteria for life, can be compared with the mode of existence of this virus in order to illustrate some of the ambiguity that exists in determining what actually constitutes "life."

The basic form of the bacteriophage has been determined by electron micrographs and chemical analysis. Structurally, this phage has a hexagonally shaped "head" attached to a thin, tubular "tail." Fibers project from the end of this tail (Fig. 23-1). Chemically, the head is filled with DNA. The rest of the phage structure is protein.

When phages are introduced into a bacterial culture, they somehow get reproduced. If the medium outside the bacterial cells is tested for the presence of phage particles at various time intervals after phages were introduced to the culture, the number of phages

Table 23-1
A Comparison of Living Characteristics for the Human and a Rock

Characteristic	Man	Rock
Complex, integrated energy requiring chemical system	yes	no
Respiration	yes	no
Growth by chemical synthesis	yes	no
Movement	yes	no
Speciation	yes	no
Responsiveness (other than gravity)	yes	no
Reproduction	yes	no

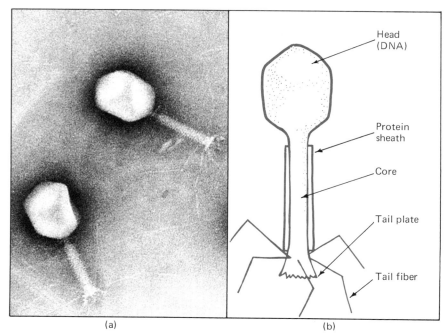

23-1 The structure of the T_4 bacteriophage particle, as shown by the electron microscope and chemical analyses. The parts of the micrograph of the phage particle (a) are given in the diagram (b). (*Courtesy of E. Wellenberger, micrograph by E. Boy de la Tour, University of Geneva, Institute of Molecular Biology; from "Vegetative Bacteriophage and Maturation of Virus Particles,"* Advan. Virus Res., **8**, 1–61 (0000). *Academic Press.*)

is constant for a certain period after their introduction to the medium. In time, however, the number of phage particles increases sharply (Fig. 23-2). This period is called the burst period. The number of phage particles within the bacterial cells can also be monitored. No phage particles, however, are found within the bacteria immediately after the phages were injected into the bacteria culture. After a time, however, the number of phages inside the bacteria also increases sharply (Fig. 23-3).

An entire phage, then, apparently does not invade the bacterial cell. In order to determine what, if any, chemical components of the phage entered the bacterium, radioactive tracers have been used. A radioactive tracer is an isotope of a particular atom which radioactively decomposes to give off either α, β, or γ rays. This decomposition can be sensed with a radioactivity detector. Radioactive atoms react chemically in a way identical to their nonradioactive isotopes. They are, therefore, incorporated into the chemical components of the new viruses being formed during viral reproduction in a bacterial culture which contains these radioactive atoms. The whereabouts of the compounds containing the radioactive atoms, however, can be followed with a radioactivity detector. Phosphorus-32 is a radioactive isotope of the nonradioactive ^{31}P which is found in the phosphate linkages of viral DNA. If ^{32}P is introduced into a bacterial medium to which phages are added, the ^{32}P gets incorporated into the DNA of the phages that are formed. The protein of the phage does not contain phosphorous, so it does not get "labelled" with ^{32}P. When these radioactive phages are placed in a bacteria culture that contains no ^{32}P, in time almost all the radioactive ^{32}P is detected *inside* the bacterial cell. The viral DNA, therefore, somehow gets injected into the bacteria.

The protein coat of the phage can also be selectively traced using radioactive sulfur, ^{35}S. When ^{35}S is introduced into a bacteria culture along with the phage, the ^{35}S is incorporated only into the protein sheaths of the phage particles that are formed. The phage DNA contains no sulfur, so it does not get labelled. When these ^{35}S-labelled phages are introduced into another bacterial culture which does not contain ^{35}S, none of the ^{35}S is found inside the

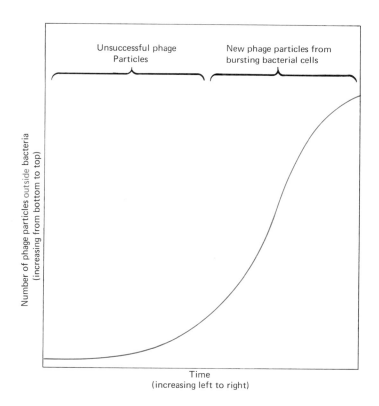

23-2

A very diagramatic representation of the number of phage particles outside bacterial cells in a new culture. The sudden increase at right comes when the inoculated bacterial cells burst and release the newly produced phage particles.

23-3

A very diagramatic representation of the number of phage particles inside bacterial cells in a new culture. The sudden increase at right comes only after sufficient time has elapsed to allow bacteriophage DNA to direct cells to form new phage particles.

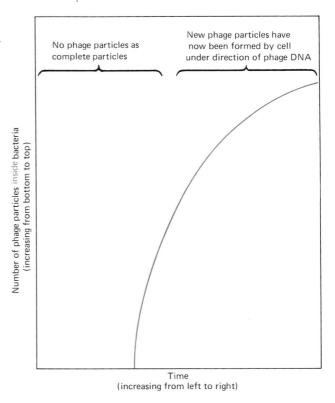

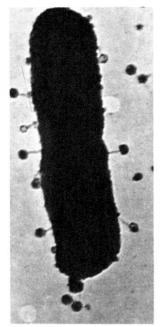

bacterial cells. This finding indicates that the protein sheath does not enter the bacterial cell. Evidently, only the viral DNA invades the bacteria.

The total picture of viral infection of the bacterial cell, based on this experimental evidence and electron microscopy, indicates that when the virus comes into contact with a suitable host, it attaches itself by the tail to this cell. It then "shoots" the DNA from the head of the virus through the tail and into the host (Fig. 23-4). The protein covering of the phage, called the ghost, stays behind and does not enter the bacterial cell. It either stays attached to the outside of the cell or floats off into the medium. Apparently this DNA that enters the bacteria is able to direct the bacterium to synthesize virus particles, and phages start to appear within the

23-4 A bacterial cell with bacteriophage particles attached to its wall by their tails. This is the position in which the bacteriophage injects its DNA into the bacterial cell. (*Photomicrograph by T. F. Anderson, The Institute for Cancer Research, Philadelphia; from "The Morphology and Osmotic Properties of Bacteriophage Systems,"* Cold Spring Harbor Symp. Quant. Biol., *18,* 197–203 (1953), *fig. 3.*)

bacteria. At some point there are so many phages present within the bacterial cell that it ruptures and releases these particles into the medium. This causes the "burst period." The phage particle is therefore *produced* by the bacterial cell. It was not present in the cell initially, and it was only present there after sufficient time had elapsed for the DNA of the phage to direct the bacterial cell components to produce phages. The total "life cycle" of this phage is summarized in Fig. 23-5.

With this background as to the virus's mode of existence, the question as to whether the virus is alive can now be attacked. The virus is a complex integrated chemical system which requires energy for maintenance and movement. However, it has no system of its own for obtaining this energy from the break down of organic foodstuffs. Instead, it relies on the system of the bacteria which it infects. Phage particles "grow" and "reproduce" inside a bacteria cell utilizing the synthesizing mechanisms and energy-producing mechanisms, as well as the raw materials, of the bacterial cell. This is the only way they can "grow" or "reproduce," since they have no independent mechanism of their own. There are many different kinds of viruses, so they definitely speciate. Viruses also have some responsiveness, since they can attach themselves to bacterial cells.

The virus, then, answers all the criteria for living except that it depends upon another organism to perform the necessary reactions for growth, reproduction, movement, and maintenance. Outside the bacterial cell the virus has a passive existence, it can only

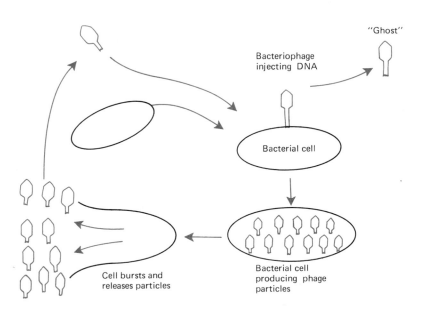

23-5

The "life cycle" of the T_5 bacteriophage.

"direct" its growth and reproduction inside its bacterial host. Whether the virus is considered "alive" is a matter of choice as to the important criteria of life. No matter what criteria are selected, however, some borderline case will no doubt put the line of demarcation in question.

What Is Science?

Science can be thought of either as a process of investigation or as the body of knowledge which has resulted from that process. The process of science has been dubbed the "scientific method." This title is somewhat misleading since there is no one "pat" method by which a scientist works. The scientific method is merely a certain way of looking at natural phenomena in order to obtain insight as to their origins, causes, and/or workings. Some specific processes which are basic to this way of looking at things are observation, measurement, and experimentation.

In order to be attacked "scientifically," a phenomenon must be observed in such a way as to pose a particular question in the observer's mind. How many apples fell before Newton started to wonder why they did? How many dwarf sweet peas appeared in the offspring of tall plants before Mendel wondered how often that happened?

Once the question is posed, one or several possible answers may come to mind. Each one is a hypothesis. The one which best fits the initial observations is tested for validity. An experiment is designed to test this proposal. Experiments can take many forms depending on the phenomenon under scrutiny. An experiment is merely a series of observations under controlled conditions. In many cases the experimental observations must be quantified so that they can be compared. As a result science is closely associated with these processes of experimentation and measurement in the minds of many people.

The results of a well-designed experiment should indicate whether the tested hypothesis is valid or not. If the hypothesis is apparently valid, then it is tentatively accepted. If it was invalid, the experimental results may point to another hypothesis which must be subjected to experimental testing. Basic to the scientific method is the requirement for evidence to corroborate a hypothesis before it is accepted. On the other hand, no hypothesis is ever conclusively proved in science; it is merely accepted until a further observation contradicts or modifies it. This objectivity is the essence of science. The scientific process must include a willingness to accept the fallibility of the body of scientific knowledge.

Science can only study phenomena that can be measured. The phenomena under scrutiny do not necessarily need to be visible, just measureable. The electron is not visible, for instance, but its properties can be measured. Moral questions, on the other hand, cannot be directly answered by science. Neither is there anything inherent in the scientific method as to the purposes to which science can be put. If science is to serve society, the direction or purpose for scientific inquiry must be supplied by the society.

Suggested Reading

STENT, C. S. *Molecular Biology of Bacterial Viruses,* W. H. Freeman, San Francisco, Calif., 1963.

Glossary

References

Index

Glossary

The following definitions are provided for the convenience of the student. These are not meant to serve as a scientific dictionary but as working definitions to be used within the context of this book.

ABORTION: The expulsion of a premature fetus.

ACID: A substance whose water solution has a pH of less than 7 and which releases H^+ ions when dissolved in water.

ACTIVE TRANSPORT: A process by which a cell moves substances in a manner that can transport materials against a concentration gradient; the process requires an expenditure of energy.

ACUTE EFFECTS: As used here, and in reference to health, either rapid or severe effects.

ADAPTABILITY: The ability of an organism to change its structure, function, or behavior to allow it better to survive in its environment.

ADAPTIVE RADIATION: The evolution of one organism to many divergent types of organisms fitted to several ecological niches or modes of existence.

ADENINE: A purine nitrogenous base that is a constituent of nucleic acids.

ADP: Adenosine diphosphate.

AEROBIC: Requiring elemental oxygen.

ALBINO: An organism with deficient pigmentation. In animals, for example, the hair is white.

ALLELES: Alternative forms of the same gene. Alleles occupy the same position on homologous chromosomes; for example, the Aa genotype represents alleles of gene "A."

ALPHA (α) RAYS: Type of atomic radiation composed of a continuous stream of alpha particles. Alpha particles contain two protons and two neutrons. They are essentially helium atom nuclei.

AMINO ACID: The structural unit of proteins, with the general form

$$\text{H}\diagdown\text{N}-\underset{\underset{\text{R}}{|}}{\overset{\overset{\text{H}}{|}}{\text{C}}}-\text{C}\diagup\overset{\text{O}}{\underset{\text{OH}}{}}$$

(—R represents some attached group).

AMP: Adenosine monophosphate.

ANAEROBIC: Not requiring elemental oxygen.

ANAPHASE: The "stage" of mitosis characterized by the movement of daughter chromosomes to the opposite poles of the cell.

ANGSTROM UNIT (Å): A unit of measurement equal to 1/10,000 of a micron.

ANTICODON: The nucleotide triplet of a tRNA unit that is coded to a complementary triplet nucleotide unit on an mRNA strand. These complementary triplets are responsible for the exact positioning of tRNA units on mRNA.

AQUATIC: Growing or living in water.

ARTIFACT: As used in biology, it means that the object or result is not there naturally but is the result of human effort; for example, the preparation of a slide may cause or leave structures resulting from the process itself and not as a part of the object being mounted.

ASEXUAL REPRODUCTION: A reproduction process that does not involve the fusion of gametes.

ASTERS: Structures found in the cytoplasm of dividing animal cells that serve as the basis for the formation of the spindle.

ATOM: The smallest unit into which a chemical element can be divided and still retain the chemical characteristics of the original element.

ATOM NUCLEUS: The central structure of an atom containing protons and neutrons. This is the central point around which the electrons occur in orbital shells.

ATOMIC NUMBER: The number of protons in the nucleus of an atom.

ATOMIC WEIGHT: The weight of one atom of an element, relative to the weight of the carbon atom which is assigned a value of 12.00.

ATP: Adenosine triphosphate.

AUTECOLOGY: The study of ecology that emphasizes the interaction between one organism, or species, and its environment.

AUTOTROPHIC: An organism capable of synthesizing its own food from simple compounds. An example of this would be the green plant.

BACTERIA: Unicellular organisms which are typically microscopic and which lack a defined nucleus and cytoplasmic structures. A few of these are disease- or decay-causing organisms.

BACTERIOPHAGE: A virus that parasitizes a bacterial cell.

BETA (β) RAYS: A form of atomic radiation which is a continuous stream of beta particles. Beta particles are electrons.

BIOCHEMICAL OXYGEN DEMAND: The amount of oxygen required for organisms to break down aerobically the organic matter in water. The more organic material in the water the greater the oxygen demand to break it down.

BIOLOGICAL COMMUNITY: All the organisms of any one given area.

BIOSPHERE: The air, land, and water at the surface of the earth which is inhabited by living things.

BOD: An abbreviation for biochemical oxygen demand.

CARBOHYDRATES: The group of organic compounds, including sugars and starch, consisting of carbon atoms to which hydrogen and oxygen atoms are attached in a $C_1:H_2:O_1$ ratio or very close to it.

CARNIVORE: An organism that lives by eating other animals.

CATALYST: A substance which speeds the rate of a chemical reaction but is not consumed during the reaction.

CELL: The basic structural unit of life. It takes the general form of a unit of protoplasm surrounded by a membrane and containing some type of nucleic acid "control center."

CELL MEMBRANE: The outermost membrane of cells. Cell membrane is synonomous with plasma membrane and plasmalemma.

CELL PLATE: A structure formed by plant cells during cytokinesis to divide the cytoplasm during cell division. It forms across the equatorial plane where the new cell walls of the daughter cells later form.

CELL WALL: The nonliving covering of a plant cell produced by the cell. It is composed chiefly of cellulose.

CENTRIOLE: A cytoplasmic organelle found in animal cells which duplicates before cell division and forms the basis for the formation of the spindle. It is not found in the cells of higher plants.

CENTROMERE: A region of the chromosome to which the spindle fiber attaches during cell division. It is also the point of attachment between chromatids.

CHEMICAL EVOLUTION: As used here, the chemical and physical processes and events that led to the development of the first cells from nonliving chemicals.

CHEMOSYNTHESIS: A process of synthesizing organic food materials using inorganic molecules as both component parts and energy sources. This is opposed to photosynthesis which uses light as an energy source for the synthesis of food materials.

CHIASMA: The crossing of two chromatids of homologous chromosomes during metaphase I of meiosis.

CHLOROPHYLL: The pigment of plant cells essential for photosynthesis. It has the ability to absorb light energy and pass it on in the form of a "high-energy" electron that can be incorporated into a carbohydrate or can serve as the basis of ATP formation through electron-transfer systems.

CHLOROPLAST: The plant cell organelle in which the chlorophyll is contained and which, because of this, is also the location of photosynthesis.

CHROMATIDS: The two daughter strands of a replicated chromosome prior to their separation during cell division.

CHROMATIN: The nucleoprotein complex of the chromosomes. This name is often also applied to the chromosomes that are fully extended in the interphase cell, as opposed to the shortened chromosomes in the dividing cell.

CHROMOSOME: A nucleoprotein strand in the cell nucleus which contains the genes of the cell in its DNA. This term is proper for either the extended form of the chromosome during interphase or the contracted form during cell division, although it is more frequently used in the latter case.

CHRONIC EFFECTS: As used here and in reference to health, low level, often cumulative effects that are often so small as to be hard to detect.

CHRONOLOGY: This refers to the arrangement of events or phenomena according to the time sequence in which they occurred.

CILIA: A "hairlike" structure of many cells projecting out from the cell and capable of wavelike movement.

CO_2 FIXATION: A process during photosynthesis in which CO_2 and other chemical factors interact to produce a carbohydrate. These reactions are part of the "dark reactions" of photosynthesis.

COACERVATE: An aggregation or droplet of organic molecules that tends to form a slight film over itself.

COAGULATION: As used here, the process of bringing fine particles or molecules into larger aggregates. This process is used in coagulation-sedimentation, for example.

CODONS: The sequence of three adjacent nucleotides that code a single amino acid.

COLLOID: A permanent suspension of fine particles or molecules in the size range of 0.1–0.001 μ.

COMPETITIVE EXCLUSION: The evolutionary process whereby organisms less well adapted to a given environment are eliminated from the environment by more successful forms.

COMPOUND: A distinct substance formed by the chemical union of two or more elements in a definite proportion.

CONSERVATION: The study of human ecology that concentrates on improving the relationship of man with his environment.

CONTAMINANT: Something added, but the emphasis is on an addition that is undesirable or harmful.

CONVERGENT EVOLUTION: The process whereby organisms of different ancestry produce similar characteristics.

COVALENT BOND: The type of chemical bond formed by the "sharing" of electrons between the component atoms.

CRISTAE: An invagination of the inner of the covering membranes of the mitochondria.

CROSSING OVER: The exchange of genetic material from one homologous chromosome to another during meiosis.

CYBERNETICS: The study of the communications and controls working within energy systems.

CYTOCHROME SYSTEMS: A series of chemicals called cytochromes which function in the electron transfer involved with both photosynthesis and respiration.

CYTOKINESIS: The process by which the cytoplasm of one cell is divided between new cells resulting from cell division.

DAUGHTER CELLS: The two cells formed by the mitotic division of one cell.

DEAMINATION: The removal of an amino group ($-NH_2$) from an organic molecule.

DECAY: The break down of organic material into simpler organic or inorganic compounds. This is usually accomplished by bacteria or fungi.

DEHYDRATION SYNTHESIS: The process of formation of new compounds by chemical bonds formed through the removal of water.

DELETION: The type of mutation resulting from the loss of a region of a chromosome or gene.

DIABETES: A disorder of carbohydrate metabolism characterized by inadequate secretion of insulin by the pancrease.

DIHYBRID CROSS: A genetic cross involving two types of genes or characters.

DIMER: A molecule composed of two identical or similar units joined together. It is composed of two similar units instead of many as for the polymer.

DIPLOID: A condition wherein each type of chromosome in a cell is represented twice. Diploid thus means two complete sets of genes.

DIVERGENT EVOLUTION: The evolutionary process whereby related forms adapt to differing environments becoming less and less similar in the process.

DNA: Deoxyribonucleic acid.

DOMINANT GENE: A gene that expresses itself in the phenotype when it is present.

DORSAL: The "top" side. This would be toward the backbone of the vertebrate forms.

ECOLOGICAL DEBT: As used here, it refers to the accrued depreciation of the environment caused by the failure of the producers and consumers of goods and services to pay the total cost of producing the goods, notably the cost of *pollution control*.

ECOLOGICAL NICHE: The particular environment and way of living of an organism.

ECOLOGY: The study of the interrelationship of an organism to its environment.

ECOSYSTEM: A functional system composed of the living organisms and the nonliving materials which interact in a given area.

EFFLUENT: As used here, the material that flows forth from society, that is, its wastes.

ELECTRODIALYSIS: A process of removing ions from water through a semipermeable membrane by the use of electrodes that are charged to "pull" the ions toward them.

ELECTROLYTES: Substances that form ions when dissolved in water.

ELECTRON MICROGRAPH: A picture taken by means of an electron microscope.

ELECTRON MICROSCOPE: A microscope whose function depends on the use of a beam of electrons rather than a beam of light. This mechanism permits magnification of up to about 300,000 diameters.

ELECTRON-TRANSFER SYSTEMS: A series of chemicals which pass electrons from one to another with a release of energy. This energy often is used in the synthesis of ATP.

ELEMENTS: Any substance composed of only one type of atom.

EMBRYONIC: Referring to the early developmental stages of an organism.

ENDERGONIC: A chemical reaction requiring energy.

ENDOPLASMIC RETICULUM: A system of membranes in the cell which serves as a point of attachment for the ribosomes and serves other functions.

ENDOTHERMIC: A heat-requiring chemical reaction.

ENVIRONMENT: The total surroundings—physical, biological, and chemical—with which an organism interacts.

ENZYMES: Protein molecules that act as catalysts in chemical reactions.

EUTROPHICATION: As used here, the process of "overfeeding" the water with minerals and organic nutrients with the resulting heavy plant growth and frequently with at least periodic reductions in dissolved oxygen which are damaging to animal life. A general upset of an aquatic ecosystem by the addition of too many nutrients.

EVOLUTION: The theory that plants and animals have their origin in preexisting types and that differences are due to modifications in successive generations. It can be considered genetically as changing gene frequencies.

EXERGONIC: An energy-yielding reaction. Fire is a simple example.

EXOTHERMIC: A heat-yielding process. Fire is a simple example.

FAD: Flavin adenine dinucleotide, an electron carrier.

$FADH_2$: The reduced form of FAD.

FATTY ACID: An organic molecule composed of a long carbon chain with its side bonds filled with hydrogen atoms and with a terminal carboxyl (—COOH) group.

FERMENTATION: The metabolic utilization of carbohydrates by organisms such as yeast without the involvement of elemental oxygen. One of the products of this process is ethyl alcohol.

FISSION: The process of splitting into smaller parts.

FLAGELLUM: A "hairlike" structure of the cell characterized by whiplike movements. Flagella are usually found one, two, or four per cell as opposed to cilia which may be numerous.

FOOD: Food is basically a concentrated form of chemical energy taken in or produced by living forms to meet their energy demands for continuing metabolism. Implied, however, is that food may have to furnish other materials necessary to the proper functioning of the organism.

FOSSIL: A remnant of a former organism. For example, coal represents the remnants of many plants which grew millions of years ago.

FRAME-SHIFT MUTATION: A mutation resulting from the insertion or deletion of a nucleotide pair. It results in a shift of the "reading frame" of the codon sequence and disrupts the triplet sequence following the point of insertion or deletion.

FUEL: Fuels are sources of concentrated energy, like food, but the energy is utilized in a nonliving process and without enzyme systems. A car, for example, burns a fuel mechanically; a human burns food enzymatically.

FUNGI: A group of heterotrophic (lacking chlorophyll) plants which are mainly parasitic or decay organisms. Bread mold and mushrooms are common examples.

GALACTOSEMIA: A disease characterized by the inability to utilize galactose, a component of milk sugar, and caused by an enzyme deficiency.

GAMETE: The functional haploid sex cell.

GAMETOPHYTE: The haploid stage in the life cycle of plants that usually produces the gametes for sexual reproduction.

GAMMA (γ) RAYS: A form of atomic radiation composed of "rays" similar to x-rays. The gamma ray is a wave form of energy rather than a particle such as is the case with alpha and beta atomic radiation.

GENE: A unit of heredity information found on a small section of a chromosome.

GENE FLOW: The process in which genes move through an interbreeding population by the mechanism of sexual reproduction.

GENE FREQUENCY: The relative occurrence of a particular gene in the gene pool of a given population.

GENE POOL: The total genetic makeup of any population. The total of all the genes within a population.

GENETICS: The study of heredity.

GENOTYPE: The genetic makeup of an organism. The genes an organism carries.

GERICIDE: The practice of killing of the aged of a population.

GLUCOSE: A sugar with six carbon atoms. This is the "blood sugar" of humans.

GLYCOLYSIS: The basic process of biochemical break down of glucose to pyruvic acid which is common to the respiratory break down of glucose, fermentation, and anaerobic production of ATP by muscles.

GOLGI APPARATUS: A system of membranes in the cell which functions in the "packaging" of certain products of the cell for internal use or secretion.

GRANA: A membranous structure located inside a chloroplast containing the pigment chlorophyll. It has a "stacked" appearance in cross section.

GUANINE: A purine nitrogenous base found in nucleic acids.

HAPLOID: The cell type in which each chromosome is represented once. This cell has only one complete set of genes as opposed to two sets in the diploid cell.

"HARD" PESTICIDES: This is applied to the pesticides that are particularly stable in the environment. Often mentioned in this respect are such common insecticides as DDT, aldrin, toxaphene, dieldrin, and chlordane.

HARDY-WEINBURG PRINCIPLE: Gene frequencies are not dependent upon dominance or recessiveness but may remain essentially unchanged in frequency from one generation to the next where no natural selection forces are acting.

HERBIVORE: An organism that lives by feeding on plants.

HETEROTROPHIC: Any organism which cannot manufacture organic compounds for its own use but must feed on organic compounds in other plants or animals.

HETEROZYGOUS: The condition where the two members of a pair of genes located on homologous chromosomes are different.

HIGHER ANIMALS: Used here to refer to animals such as the vertebrates. The general reference here is to the vertebrates.

HIGHER PLANTS: Used here to refer to plants that have multicellular specialized structures to serve in the conduction of fluids within the plant; the vascular plants, such as trees, flowers, bushes, and grasses.

HOMOLOGOUS CHROMOSOMES: The pair of chromosomes within a cell which carries genes influencing the same characteristics. One of these chromosomes is maternal in origin and the other paternal in origin.

HOMOZYGOUS: The condition in which two alleles of a gene on the homologous chromosome are identical.

HYDROCARBON: An organic compound that is composed of carbon and hydrogen.

HYPOTHESIS: A tentative explanation for an observed phenomenon.

INFANTICIDE: The practice of killing of infants.

INORGANIC COMPOUND: A compound that does not contain a chain of two or more carbon atoms linked together. Generally these are compounds that do not contain carbon, but CO_2 and H_2CO_3 are generally considered inorganic.

INSULIN: A compound essential for proper utilization of carbohydrates by the human body; it is produced in the pancrease.

INTERPHASE: The "resting" phase between cell divisions.

INVERSION: A process in which a section of a chromosome breaks away and then is reattached in an inverse order for its genes.

ION: An atom or group of atoms that carries a positive or negative charge as the result of having gained or lost electrons from its original form.

IONIC BOND: A chemical bond that has, as its bonding force, the attraction of ions with opposite charges.

IONIC BONDING: The type of chemical bonding whereby one atom either "gives" away or "accepts" an electron in order to reach a stable condition in every orbital shell. This results in the formation of ions which are attracted or bonded together by their opposite ionic charges.

KINETIC ENERGY: The energy of motion.

KREBS' CYCLE: A cycle of reactions involving the oxidation of pyruvic acid and during which large amounts of energy are released. The energy released as the result of this process (and electron transfer) is used to synthesis ATP.

LINEAR: Refers to a linelike form or arrangement.

LOCUS: As used here, the position of a gene on a chromosome.

LYSIS: The process of break down.

LYSOSOME: Cytoplasmic organelles involved in the breakdown of fats, proteins, nucleic acids, and foreign particles in the cytoplasm.

MACROMOLECULE: A molecule of very high molecular weight and formed by polymerization.

MALNUTRITION: The insufficient intake of proteins, vitamins, minerals, and so forth, to meet the needs of the body, whether or not sufficient calories are taken in.

MAMMALS: The group of vertebrates characterized by the presence of mammary glands for nursing of young and hair on the body.

MARINE: Refers to living in a saltwater environment.

MATRIX: The medium within something is located.

MEGAGAMETOPHYTE: The "plant" that produces the female sex cell or egg cell. In the higher plants this "plant" is microscopic and enclosed by the structures of the sporophyte.

MEGASPORE: The cell that develops into the female gametophyte.

MEGASPORE MOTHER CELL: The cell that gives rise to the megaspore.

MEIOSIS: The process of cell division without formation of identical cells, and which forms haploid cells with a random sampling of the maternal and paternal chromosomes of the original cell.

MESSENGER RNA (mRNA): A type of RNA and factor in the process of protein synthesis. It codes a sequence of amino acids in a specific protein by means of a linear sequence of triplet codons which are coded to tRNA units with specific amino acids attached.

METABOLISM: The total of the chemical activities of the cell.

METAPHASE: The stage of mitosis characterized by the "lining up" of the chromosomes along the equatorial plates of the cell.

METAPHASE PLATE: The region across the central portion of the cell where the chromosomes align themselves at metaphase.

MICROGAMETOPHYTE: The "plant" that produces the male sex cell or sperm cell. In the higher plants this stage is represented by the pollen grain and its fully developed form.

MICROSPORE: The cell that develops into the male gametophyte.

MILLIMICRON (mμ): A unit of measurement which is $1/1000\ \mu$ or $1/1,000,000$ mm.

MITOCHONDRIA: The cytoplasmic organelle that functions as the main site for the production of ATP through the Krebs' cycle and electron transfer.

MITOSIS: The process of cell division whereby two genetically identical daughter cells are produced.

MOLECULE: A chemical structure consisting of two or more atoms bonded together. It is a compound if the atoms are of two or more different kinds.

MONOHYBRID CROSS: A genetic cross involving only one type of genes or characters.

MONOMER: A chemical structure that can be bonded together with many of the same

Glossary [313]

or similar units to form a long chain of such units called a polymer.

MUNICIPAL WASTES: The waste products of cities. Since most Americans now live in "cities" this includes most nonindustrial wastes.

MUTATION: A change in the structure of DNA or chromosomes which results in a permanent and inheritable change in genetic information carried by the chromosome.

NAD: Nicotinamide adenine dinucleotide, an electron carrier.

NADH$_2$: Reduced form of NAD.

NADP: Nicotinamide adenine dinucleotide phosphate, an electron carrier.

NADPH$_2$: Reduced form of NADP.

NATURAL SELECTION: The evolutionary process whereby organisms with hereditary characteristics favorable to that environment survive and reproduce their kind faster than those with less favorable characteristics.

NEGATIVE FEEDBACK: The process whereby the product of a functioning system feeds back into the system to decrease its own production.

NEUTRON: A subatomic particle without an electrochemical charge.

NITROGEN OXIDES: Compounds of oxygen and nitrogen, most notably nitrous oxide (N_2O), nitric oxide (NO), and nitrogen dioxide (NO_2).

NITROGENOUS BASE: A purine or pyrimidine—a nitrogen containing molecule—with basic properties.

NOISE: Unwelcome sound. Sound that is music to some is "noise" to others.

NOTOCHORD: The cartilaginous rod dorsally located and found either embryonically or in the mature animal for all chordates.

NUCLEAR ENVELOPE: The outmost portion of the nucleus, which consists of two membranes containing pores. It is not currently known whether these pores are open or themselves have a fine covering.

NUCLEOLUS: A spherical body within the nucleus, probably the site of ribosomal RNA synthesis.

NUCLEOPLASM: The living material within the nucleus.

NUCLEOPROTEIN: The complex chemical containing both nucleic acids and proteins.

NUCLEOTIDE: The "building block" of nucleic acids; formed from one unit of nitrogenous base, one unit of sugar, and one phosphate group.

NUCLEUS: *Biological:* The spherical body found in most cells which directs cell function and contains the hereditary material of the cell. *Chemical:* The central core of the atom which contains the protons and neutrons.

OLEFIN: A group of organic compounds that contain at least one double carbon-to-carbon bond.

OMNIVORE: An organism that eats both plants and animals.

OPERON: A group of adjacent genes with its operator gene, which together make up one functional unit.

ORBITAL SHELLS: The series of (energy) levels extending out from the nucleus of an atom within which the electrons occur.

ORGANELLES: Small structures of varying size and function appearing within the cytoplasm of the cell.

ORGANIC COMPOUNDS: Compounds that contain two or more linked carbon atoms, and some compounds containing single or nonlinked carbon atoms. Methane and formaldehyde, for example, are generally considered organic although containing only a single carbon atom.

ORGANISM: An individual living creature.

OXIDATION: The chemical process involving the removal of electrons. An example of oxidation is the loss of an electron by an excited chlorophyll molecule to go from Chl to Chl$^+$ and e^-.

OXYGEN REVOLUTION: A term applied to a series of chemical reactions that apparently occurred early in the history of the earth as a result of elemental oxygen being made available through the process of photosynthesis.

OZONE: A molecule composed of three oxygen atoms.

PALEOBOTANY: The study of ancient and fossil plants.

PALEOZOOLOGY: The study of ancient and fossil animals.

PANCREAS: A large organ of the vertebrates that secretes digestive enzymes and the hormone insulin.

PARASITE: An organism that receives its food supply from the body of another organism, usually without killing it directly.

PARTICULATE CONTAMINANTS: Contaminants that are at the molecular aggregate, rather than the molecular, level. In air and water pollution the term usually applies to particles that are relatively inert chemically.

PEPTIDE: The structure formed by the linking of two or more amino acids.

PETROLEUM LIQUIDS: This term includes both crude oil and the liquids which are often recovered with natural gas. Propane and butane are two such "natural gas liquids."

PGA: Phosphoglyceric acid.

PGAld: Phosphoglyceraldehyde.

PHENOTYPE: The outward appearance of an organism, as contrasted with its genetic makeup.

PHENYLKETONURIA: A disease characterized by severe feeblemindedness and caused by an enzyme deficiency that prevents utilization of the common amino acid phenylalanine.

PHOSPHORYLATION: The addition of a phosphate group to a compound.

PHOTOCHEMICAL SMOG: A complex of air contaminants that are formed by the action of sunlight on smog.

PHOTOLYSIS: The splitting of a compound by involvement of light energy.

PHOTOSYNTHESIS: The process by which carbohydrates are manufactured from carbon dioxide and water by chlorophyll-containing protoplasm using light as an energy source.

PHYLOGENY: The study of the evolutionary relationships between organisms.

PHYLUM CHORDATA: The taxonomic group characterized by dorsal nerve cord, a notochord present either in the embryo or mature form, and pharyngeal gill slits.

PLASMA MEMBRANE: The thin membrane covering the outside of the cell.

POINT MUTATION: A mutation involving a single nucleotide pair of the DNA molecule.

POLARITY: A condition where small electrochemical charges develop at specific locations on molecules as the result of slightly uneven frequency distribution of the electrons of the structure.

POLLUTION: Contamination that has undesirable or harmful effects. Something polluted is something damaged by contamination.

POLYMER: A very large molecule composed of many similar molecular units linked together. Starches, cellulose, proteins, and nucleic acids are examples.

POLYSOME: The complex of an mRNA strand and several ribosomes attached to it at separate points that are reading off its codons.

POPULATION DYNAMICS: The sum total of the interactions of a population with its environment. This emphasizes the relation of the population to its environment, especially with regard to birth and death within its population, and its energy relationships with the environment.

POSITIVE FEEDBACK: A condition whereby the product of a functioning system feeds back into the system to increase its production.

PRECIPITATION: This is water falling upon the earth in the form of rain, snow, sleet, or hail.

PREDATOR: An animal that obtains its food supply by direct killing of other animals.

PRIMATE: The order containing man, the apes, monkeys, and related forms.

PROPHASE: The stage of mitosis marked by the visible formation of chromosomes, the disappearance of the nuclear envelope, and the formation of the spindle.

PROTEIN: An organic compound made up of many amino acids linked by peptide bonds with a molecular weight of at least 30,000.

PROTON: A subatomic particle with a positive charge.

PROTOPLASM: The living material of all cells.

PROTOZOA: Most generally applied to small, unicellular animals.

PUNNETT SQUARE: A form of chart used for calculation of all possible combinations of gametes formed by meiosis. A method of working elementary genetics problems.

PURINES: Nitrogen-containing compounds called nitrogenous bases. The structure consists of two fused rings containing carbon and nitrogen atoms, and varying side groups.

PYRIMIDINES: Nitrogen-containing compounds called nitrogenous bases. The structure consists of a single ring containing carbon and nitrogen atoms and varying side groups.

RADIATION: A process involving the movement of energy. Sunlight, x-rays, and atomic radiation are examples.

RADIOACTIVE COMPOUNDS: Substances that emit radiation because of changes in their unstable nuclei. Although stable for relatively long periods of time they eventually change to other compounds through nuclear reactions.

RADIOACTIVE DATING: A process of measurement of age by means of calculation of amounts of radioactive material in mineral deposits or organisms.

RAYON: Reconstituted cellulose, usually in the form of threads or cords.

RECESSIVE GENE: An allele whose phenotypic expression is masked by the presence of a dominant. A gene that expresses itself only in the absence of a dominant gene.

REDUCTION: The chemical process involving the gain of electrons. As used here it is most often applied to electron carriers such as NAD or FAD. For example, NAD accepts electrons (with accompanying protons) and is reduced from NAD to $NADH_2$.

REPRODUCTIVE POTENTIAL: The total capacity of a population to increase in the absence of control factors. The theoretical increase rate.

RESPIRATION: The break down and release of energy from food-fuel compounds through reactions requiring elemental oxygen.

RESPONSIVENESS: The capacity to respond to environmental conditions.

REVERSE OSMOSIS: A process forcing water to diffuse through a semipermeable membrane against a concentration gradient of solutes by the application of pressure to the solution containing the dissolved materials.

RIBONUCLEIC ACID (RNA): The nucleic acid involved in protein synthesis. It is so named because its sugar is ribose.

RIBOSE: A sugar containing five carbon atoms.

RIBOSOMES: Particles found in the cytoplasm or attached to the endoplasmic reticulum. The site of protein synthesis.

RNA: Ribonucleic acid.

rRNA: A type of RNA found on the ribosomes.

SCIENTIFIC METHOD: The process of observation, hypothesis, controlled experimentation, and conclusion resulting in verifiable data and facts.

SEDIMENTATION: The process of material settling out of a liquid. The materials settling out are called sediment.

SEWAGE: The wastes of society—solid, suspended, or dissolved—when carried by liquid through the sewer.

SEXUAL REPRODUCTION: The type of reproduction involving the fusion of two haploid gametes, one from the male and one from the female parent.

SMOG: Originally a combination of fog with smoke to give heavy air contamination; it is now used to refer to visible air pollution.

SOCIAL DEBT: As used here, it refers to the cost, inconvenience, and problems caused to society in general by the failure of producers and consumers to pay the total cost of goods and services, notably various forms of pollution control.

SOLUTE: A material that dissolves in a solvent. Table salt, the solute, dissolves in water, the solvent.

SOLUTION: A liquid containing a dissolved substance. This substance is present at the ionic or molecular level.

SOLVENT: A material that dissolves another material. Water, the solvent, dissolves table salt, the solute.

SPECIES: A term referring to a population which is recognizably and genetically distinct from other groups.

SPINDLE FIBER: A "threadlike" structure formed during the prophase stage of mitosis, radiating out from the asters.

SPONTANEOUS: Something spontaneous occurs without influences. It is self-generated.

SPOROPHYTE: The stage in the life cycle of the plant that produces spores.

STARCH: A complex carbohydrate. It is a polymer of glucose.

SUBSCRIPT: As used here, it is a number written below the line of writing, such as to indicate the atomic number of an atom, $_2$He, or the number of atoms in a compound, H_2O.

SUGARS: Carbohydrates that usually contain five or six carbon atoms, or combinations of two such five- or six-atom units. Dextrose is a six-carbon unit and sucrose is a combination of two six-carbon units.

SULFUR OXIDES: Compounds of sulfur and oxygen; commonly either sulfur dioxide (SO_2) or sulfur trioxide (SO_3).

SYNECOLOGY: The study of ecology that emphasizes the interaction of whole aggregates of organisms with their environment.

TAXONOMY: The study of the classification of organisms. This involves the study both of identification of animals and plants and of their relationships.

TELOPHASE: The last stage of mitosis during which two daughter nuclei are formed.

TEMPERATURE INVERSION: A weather condition that traps air contamination near the ground with a "lid" of warm air which blocks normal air convection currents.

TEMPLATE REPLICATION: A process of replication that involves a form or pattern as its basis.

THEORY: A proposed explanation based on observation and controlled experimentation.

THERMODYNAMICS: The study of energy systems and energy flow.

THYMINE: A pyrimidine nitrogenous base and constituent of nucleic acid.

TOXIC: Poisonous.

TRANSCRIPTION: A term applied to the process of RNA formation from a DNA template.

TRANSFER-RNA (tRNA): A type of RNA and factor in the process of protein synthesis. It is responsible for the very selective attachment to amino acids in the cytoplasm and the precise positioning of these amino acids on the ribosome-mRNA complex so that they will have the proper position in the resulting protein.

TRANSITIONAL MUTATION: A point mutation involving the changing of a nucleotide pair through a series of steps including replication.

TRANSLOCATION: The transfer of chromosome material from one chromosome to a nonhomologous chromosome by breakage and reattachment.

TRIPHOSPHODEOXYRIBONUCLEOTIDE: A deoxyribonucleotide with a triple phosphate chain attached to the deoxyribose unit where a single phosphate would normally be attached.

UNDERNUTRITION: The intake of insufficient

calories to meet the needs of the body.

URACIL: A pyrimidine nitrogenous base and component of RNA.

VACUOLE: A saclike structure in the cytoplasm often containing waste materials or food.

VERTEBRATES: Animals with a (vertebral) backbone.

VIRUS: A parasitic, noncellular, particle composed of a protein shell around a nucleic acid core. It is capable of reproduction only within a host cell.

ZYGOTE: The single diploid cell resulting from the fusion of egg and sperm cells.

References

AARONSON, TERRI. "Tempest Over a Teapot," *Environment*, **11**(8), 22–27 (1970).

Agricultural Statistics 1969. U.S. Department of Agriculture, U.S. Government Printing Office, Washington, D.C.

Air Pollution Primer. National Tuberculosis and Respiratory Disease Association, 1740 Broadway, New York, N.Y., 10019, 1969.

ALLEN, S. W., and J. W. LEONARD. *Conserving Natural Resources*, McGraw-Hill Book Company, New York, 1966.

A Primer on Air Pollution. 2nd ed., Mobil Oil Corporation, 150 East 42nd Street, New York, N.Y., 10017, 1970.

A Primer on Waste Water Treatment. Document of the U.S. Department of the Interior, Federal Water Pollution Control Administration, 1969.

BENARDE, MELVIN A. *Our Precarious Habitat*, W. W. Norton, New York, 1970.

BENNETT, T. P., and EARL FRIEDEN. *Modern Topics in Biochemistry*, Macmillan, New York, 1966.

BORGSTROM, GEORG. *The Hungry Planet*, Collier-Macmillan, London, 1967.

BORGSTROM, GEORG. *Too Many*, Collier-Macmillan, London, 1969.

BROOKS, R. R. "People Versus Food," *Saturday Rev*, September 5, 1970 (pp. 10–14, 33).

"By-Products Await Desalter." *Chem. Eng. News*, Ind. Bus. Sect., July 13, 1970 (pp. 18–20).

CADLE, R. D., and E. R. ALLEN. "Atmosphere Photochemistry," *Science*, **167**, 243–249 (1970).

"City vs. Forest." Editorial in *Time Mag.*, April 13, 1970 (p. 49).

Cleaning Our Environment: The Chemical Basis for Action. A report by the Subcommittee on Environmental Improvement, Committee on Chemistry and Public Affairs of the American Chemical Society, 1969.

Comparative Emissions from Some Leaded and Prototype Lead-Free Automobile Fuels. Rep. Invest. 7390, Bureau of Mines, United States Department of the Interior, 1970.

Control of Automobile Emissions. Technical Information Service, Public Relations Staff, Ford Motor Company, Dearborn, Michigan, 1969.

CRAM, D. J., and G. S. HAMMOND. *Organic Chemistry*, McGraw-Hill, New York, 1964.

CRICK, F. H. C. "The Structure of Hereditary Material," *Sci. Amer.* **191** (4), 54–61 (1954).

CROW, J. F. *Genetics Notes*. Burgess Publishing Company, Minneapolis, Minn., (1960).

DARLEY, E. F., C. W. NICHOLS, J. T. MIDDLETON. "Identification of Air Pollution Damage to Agricultural Crops," *The Bulletin*, Department of Agriculture, State of California, **55**, 11–19 (1966).

DARWIN, CHARLES. *The Origin of Species by Means of Natural Selection or the Preservation of Favored Races in the Struggle for Life,* (original publication 1859), Oxford University Press, New York, 1951.

DASMAN, R. F. *Environmental Conservation,* John Wiley & Sons, New York, 1968.

DEBUSK, A. G. *Molecular Genetics,* Macmillan, New York, 1968.

DODSON, E. O. *Evolution: Process and Product,* Reinhold Publishing Company, New York, 1960.

EDWARDS, C. A. "Insecticide Residues in Soils," *Residue Rev.,* **13,** 83–132 (1966).

EHRENFELD, D. W. *Biological Conservation,* Holt, Rinehart and Winston, New York, 1970.

EHRLICH, P. R., and A. H. EHRLICH. *Population, Resources, Environment,* W. H. Freeman, San Francisco, Calif., 1970.

EHRLICH, P. R., and A. H. EHRLICH. "The Food-From-The-Sea Myth," *Saturday Rev.,* April 4, 1970 (pp. 53–65).

FAWCETT, D. W. *An Atlas of Fine Structure, The Cell, Its Organelles, and Inclusion,* Saunders; Philadelphia, Pa., 1966.

Final Report on Industrial By-Product Recovery by Desalination Techniques. Prepared by Aqua-Chem Incorporated, Waukesha, Wisconsin, for the Department of the Interior, Office of Saline Water, 1970.

GAMOV, G. "The Origin and Evolution of the Earth," *Amer. Sci.,* **39,** 393–406 (1951).

GIESE, A. C. *Cell Physiology,* Saunders, Philadelphia, Pa., 1968.

GOFMAN, J. W., and ARTHUR TAMPLIN. "Radiation: The Invisible Casualties," *Environment,* **12** (3), 11–19, 49 (1970).

GOLDSBY, R. A. *Cells and Energy,* Macmillan, New York, 1967.

GRAHAME, ARTHUR. "Last Chance for Atlantic Salmon?" *Outdoor Life,* June, 1970 (pp. 41–43, 92, 94, 96, 98, 100).

GRANT, NEVILLE. "The Legacy of the Mad Hatter," *Environment* 11 (4), 18–23, 43–44 (1969).

GUYTON, A. C. *The Function of the Human Body,* Saunders, Philadelphia, Pa., 1965.

HOLMES, ARTHUR. *Principles of Physical Geology,* Ronald Press, New York, 1965.

JACOB, F. and J. MONOD. "Genetic Regulatory Mechanisms in the Synthesis of Protein," *J. Mol. Biol.,* **3,** 318–356 (1961).

JELLINCK, P. H. *The Cellular Role of Macromolecules,* Scott, Foresman and Company, Glenview, Ill., 1965.

Keep It Clean. Highlights of Bethlehem's Pollution Control Program, Bethlehem Steel Corporation, Bethlehem, Pa., 18016, 1970.

LEAR, JOHN. "The Crisis in Water: What Brought It On?" *Saturday Rev.,* October 23, 1965 (pp. 24–28, 78–80).

LEAR, JOHN. "Green Light for the Smogless Car," *Saturday Rev.,* December 6, 1969 (pp. 81–86).

LEAR, JOHN. "A Progress Report on Smogless Motoring," *Saturday Rev.,* August 1, 1970 (pp. 44–45).

LÖFROTH, GÖRAN, and M. E. DUFFY. "Birds Give Us Warning," *Environment,* **11**(4), 10–17 (1969).

MARTELL, E. A. et al. "Fire Damage," *Environment,* **12**(4), 14–21 (1970).

MARX, WESLEY. 1967. *The Frail Ocean.* Ballantine: New York.

"Menace in the Skies." Editorial in *Time Mag.,* January 27, 1967 (pp. 48–52).

MENDEL, G. "Experiments in Plant Hybridization," *Verhandlungen naturforschender Verein in Brunn Abhundlungen,* **iv** (translations available).

MILLER, P. R. et al. "Ozone Dosage Response of Ponderosa Pine Seedlings," *APCA J.,* **19,** 435–438 (1969).

MILLER, S. L. "A Production of Amino Acids under Possible Primitive Earth Conditions," *Science,* **117,** 528–529 (1953).

Minerals Yearbook. U.S. Department of the Interior, U.S. Bureau of Mines, 1968.

MORTIMER, C. E. *Chemistry, A Conceptual Approach,* Reinhold, New York, 1967.

MURPHY, D. B., and VIATEUR ROUSSEAU. *Fundamentals of College Chemistry,* Ronald Press Company, New York, 1969.

NASH, R. G., and E. A. WOOLSON. "Persistance of Chlorinated Hydrocarbon Insecticides in Soils, *Science,* **157,** 924–927 (1957).

NOVICK, SHELDON. "A New Pollution Problem," *Environment,* **11**(4), 3–9 (1969).

OPARIN, A. I. *Life, Its Origin, Nature, and Development,* Ann Synge (trans.), Academic Press, New York, 1966.

"Pollution Price Tag: 71 Billion Dollars." *U.S. News World Rep.,* August 17, 1970 (pp. 38–42).

RASMUSSEN, D. I. "Biotic Communities of the Kaibab Plateau," *Ecol. Monog.,* **3**, 229–275 (1941).

Resources and Man. Committee on Resources and Man, National Academy of Sciences, National Research Council, W. H. Freeman, San Francisco, Calif., 1969.

RIENOW, ROBERT, and L. T. RIENOW. *Moment in the Sun,* Ballantine, New York, 1967.

RYTHER, J. H. "Photosynthesis and Fish Production in the Sea," *Science,* **166**, 72–76 (1969).

SAVAGE, J. A. *Evolution,* Holt, Reinhart, Winston, New York, 1963.

SLEIGH, N. A. *The Biology of Cilia and Flagella,* Pergamon Press, Elmsford, N.Y., 1963.

SMITH, GUY-HOWARD (ed.). *Conservation of Natural Resources,* John Wiley & Sons, New York, 1965.

SORUM, C. H. *Fundamentals of General Chemistry,* Prentice-Hall, Englewood Cliffs, N.J., 1955.

STANIER, R. Y., M. DOUDOROFF, and A. E. ALDELBERG. *The Microbial World,* Prentice Hall, New York, 1964.

STENT, G. S. *Molecular Biology of Bacterial Viruses,* W. H. Freeman, San Francisco, Calif., 1963.

STRICKBERGER, N. W. *Genetics,* Macmillan, New York, 1968.

Summary of Registered Agricultural Pesticide Chemical Uses. 3rd ed., vol. III, Pesticides Regulation Division, Agricultural Research Service, U.S. Department of Agriculture, Washington, D.C., May 31, 1969.

TAYLOR, G. R. "The Threat to Life in the Sea," *Saturday Rev.,* August 1, 1970 (pp. 40–42).

TAYLOR, O. CLIFTON. "Effects of Oxidant Air Pollutants," *J. Occup. Med.,* **10**, 485–492 (1968).

TAYLOR, O. CLIFTON. "Importance of Peroxyacetyl Nitrate (PAN) as a Phytotoxic Air Pollutant," *APCA J.,* **19**, 347–351 (1969).

TAYLOR, O. CLIFTON. "Agriculture and Air Pollution," *Calif. Air Environ.,* **1**, 1–3 (1970).

The American Heritage Dictionary of the English Language. American Heritage Publishing Co., New York, and Houghton Mifflin Company, New York, 1969.

The Demand and Price Situation for Forest Products 1968–1969. U.S. Department of Agriculture, Forest Service, Miscellaneous Publication no. 1086, 1969.

The Living Cell. Readings from *Scientific American,* W. H. Freeman, San Francisco, Calif., 1965.

THOMPSON, C. R., et al. "Effects of Air Pollutants on Apparent Photosynthesis and Water Use by Citrus Trees," *Environ. Sci. Technol.,* **1**, 644–650 (1967).

THOMPSON, C. R., E. HENSEL, and G. KATS. "Effects of Photochemical Air Pollutants on Zinfandel Grapes," *Hort. Sci.,* **4**, 222–224 (1969).

THOMPSON, C. R., and O. C. TAYLOR. "Effects of Air Pollutants on Growth, Leaf Drop, Fruit Drop, and Yield of Citrus Trees," *Environ. Sci. Technol.,* **3**, 934–940 (1969).

THOMPSON, C. R. "Effects of Air Pollutants in the Los Angeles Basin on Citrus," *Proc. 1st Int. Citrus Symp.,* **2**, 705–709 (1969).

Timber Trends in the United States. U.S. Department of Agriculture, Forest Service, Res. Rep. no. 17, 1965.

TRIPPENSEE, R. E. *Wildlife Management, Upland Game and General Principles,* McGraw-Hill, New York, 1948.

UREY, H. "The Origin of the Earth," *Sci. Amer.,* **187**(4), 53–60 (1952).

WALD, G. "The Origin of Life," *Sci. Amer.,* **191**(2), 44–53 (1954).

WEINBERG, A. M., and R. P. HAMMOND. "Limits to the Use of Energy," *Amer. Sci.,* **58**, 412–418 (1970).

WURSTER, C. F. "DDT Reduces Photosynthesis by Marine Phytoplankdon," *Science,* **159**, 1474–1475 (1968).

ZEUNER, F. E. *Dating the Past, An Introduction to Geochronology,* Metheun & Company, London, 1946.

Index*

Abortion, 235
Acetaldehyde, 104, 105, **105**
Acetic Acid, 26
Acetyl-coenzyme A (complex), 102, **102,** 103, 107, 108
Acid, 50, 51, 52, 84, 86, 258, 259
Acrolein, 245, 246, **246,** 247
Active transport, 110, *114,* **115,** 116, 218, 219, 220, 227
Activated amino acid, 80
Activated sludge, 262
Activator enzyme, 80
Adaptive radiation, *186,* 187, 194
Adenine, **19,** 56, **57,** 58, 59, **59, 63,** 69, 70, 71, 72, 76, 80, **127,** 129
Adenine nucleotide, 84, **86**
Adenosine diphosphate (ADP), 51, *56,* 58 95, 96, 98–103, 115
Adenosine monophosphate, *56,* 58
Adenosine phosphates, *56*
Adenosine ribophosphate, **19**
Adenosine triphosphate (ATP) 50, *56,* **57,** 58, 80, 94–108, 115, 206, 296
ADP (adenosine diphosphate), 51, *56,* 58, 95, 96, 98–103, 115
ATP production from general food compounds, *106*
Aeration, 262, 265

Aeration tank, 262, **262**
Aerobic breakdown of foods, 99, 256, 296
Age of earth, 196
Air, *240,* 241, 243, 244, 248, 255, 256, 262, 267, 289
Air contaminants, 240, 241, *242,* 245, 247, 248, 250, 252
Air contaminant levels, *249*
Air contamination effects, *242*
Air contamination quantities, *242*
Air contamination sources, *242*
Air convection, 248
Air pollution, 242, 247, 248, 249, 250, 251, 252, 253
Air pollution and respiratory illness, *247*
Air self-purification, 241, **241**
Albino, 87
Alcohol, 17, **17,** 51, **52**
Alcoholic dehydrogenase, 105
Algae, 196, 258, 275
Alleles, 123, 150, 155, 157, 173, 174, 175, 182
Aluminum, 287, 288
Ambiguous code system, *72*
Amide, 51, **52,** 53, 54, 285
Amine, 51, **52,** 53, 54
Amino acid, 15, 17, *18,* **19,** 20, 21, 53, **53,** 54, 67, 69, 72–75, 79–84, 87, 108, 229
Amino acids, of organisms, *74*
Amino group, 18, 20, 127

*Italic numbers refer to main text discussion, **boldface** numbers refer to illustrations.

Ammonia, 12, **13**, 15, 257, 258
Amphibian, 186
Anaerobic ATP production by muscles, 100, *105*
Anaerobic bacteria, 258
Anaerobic breakdown of foods, 99, 104, 105, 257
Analogous structure, *194*
Anaphase, of mitosis, 131, *135*, **136**, 141
Anaphase I, of meiosis, *141*, **141**, 150
Anaphase II, of meiosis, *142*, **143**
Angstrom (Å), 30, 31, 34
Angora wool, 284
Animal, 121, 123, 130, 131, 135, 138, 140, 144, 145, 147, 185, 213, 215, 218, 224, 226, 227, 228, 230, 233, 234, 246, 247, 256, 257, 261, 271, 272, 273, 274, 281, 284
Animal hair products, *284*
Animal population curve, 279, **280**, 281
Animal population dynamics, *215*
Animal products, for clothing, *283*
Animals, warm blooded, 283
Anticodon, **79**, 80, 81, **82**
Antiknock agent, 245
Aquatic organisms, 256–260
Aromatic compounds, 243, 244, **244**
Artifact, 30
Asphalt, 289
Aster, 45, 131, **133, 134, 136,** 140, **140**
ATP (adenosine triphosphate), 50, *56*, **57**, 58, 80, 94–108, 115, 206, 296
Atlantic salmon, 274, 275
Atmosphere, 14, 15, 27, 239–244, 247, 248, 249, 252, 277
Atom splitter (human social development), 228, *229*, **230**
Atomic weight, 5, 7, 8
Atomic number, 4, 5, 6, 7, 9
Atomic movement (early earth), 7
Atomic-nuclear energy, 228, 231, 232, 250, 251
Atoms, structure of, 4
Automobile, 242, 243, 245, 252, 253
Autotroph, *26*, 27
Average cell, **46**
Bacteria, 30, 91, 117, 121, 130, 203, 219, 220, 257, 262, 263, 297–302, **300**

Bacteriophage, *297*, **298**, 299, **300**
Basal plate, 43
Base, 50, 86, 258
Base pairing, 75
Behavioral isolation, 182
Benzene, 244, **244**, 276
Bicarbonate (ion), 50, 114
Biochemical oxygen demand (BOD), *256*, 257
Biology, 3, *295*
Biological community, 93, 97, 171, 185, 186, 200, 202, 203, 206, 207, 208, 212, 213, 217, 221, 223, 226, 231, 232, 239, 240, 250, 260, 261, 268
Biological concentration, *260*, 275
Biological energy needs of man, 226, 231, **232**
Biological evolution, 68, 169, 185, 196
Biological man, 232
Biological system, 200, 203, 206, 208, 211, 212, 250, 258
Biologically useable energy, 99
Biomass, 221, 222
Biosphere, 207, 231, 232, 239, 240, 247, 281, 291
Birth control, 234, *235*, 236, 281
Birth control pill, 235
Bird, 186, **190**, 194, 196, 259
Binding site, on ribosome, 81, **82**
Black plague, 233
BOD (biochemical oxygen demand), *256*, 257
Body temperature, 202, 212, 213
Bonding electron, 31, 32
Bone cell, **47**, 48
Brain, 212, 213
Breakage, of chromosome, 124, 125, 126
Breeder (nuclear) reactor, *278*, 279
Brick, 231, 285, 288, 289
Bronchitis, and air pollution, 247
Bubonic plague, 233
Burst period, of bacteriophage, 298, **299**, 301, **301**
Calcium, 50, 115
Calcium oxide, in plaster, 287
Calorie, 270–273

Carbohydrate, 20, 21, 41, 51, *55*, 58, 64, 107, 108, 110, 256
Carbon adsorption treatment (of water), 264, 266
Carbon chain, 16, **16**, 17, 18
Carbon dioxide, 12, **13**, 15, 26, 27, 50, 93, 94, 97, 98, 99, 102, 103, 104, 108, 114, 205, 206, 217, 218, 240, 241, 243, 256, 257, 276
Carbon dioxide (CO_2) fixation, 94, *97*, **97**
Carbon, important role of, *15*
Carbon monoxide, 242, 243
Carbonic acid (H_2CO_3), 50, 114
Carboxyl group, 18, 51, **52**, 53, **53**, 54, 55, 106, 107
Carnivore, 185, 202, 203, 206, 207, 217, 220, 221, 222, 228, 229, 232, 233
Carrying capacity, *170*, 171, 215, 229, 280, 281, 289, 290
Catalysis, 84, 85, 86, 88, 90
Cell, 3, **25**, 29-31, 33, 35, 37, 45, 49, 50, 51, 53, 54, 58, 59, 67, 78, 86, 88, 92, 99, 110, 114, 116, 117, 122, 123, 130, 132, **133**, **134**, 135, **136**, 139, **144**, 145, 150, 169, 196, 217, 218, 219, 298, 299, 300, 301
Cell division, 35, 45, 121, 122, 124, *130*, 131, 132, 135, 145, 147, 218, 219, 220, 227
Cell membrane, *31*, 33, 53, 55, 81, 110, 114, 115, 116, 117, 131, **133**, **134**, **136**, **137**, **138**
Cellulose, 30, 45, 55, 218, 284
Cell wall, 45, 46, 55, **132**, **133**, **134**, 135, **136**, **137**
Centriole, 29, *41*, 43, **43**, **44**, 45, 81, 131, **139**, 140
Centromere, 122, **123**, 125, 131, **133**, 135
Cesium-137, 250
Charged t-RNA, **80**, **82**, **83**
Chemical bond, *8*, 9, 20, 26, 94, 199, 202, 205
Chemical code, 69, 296
Chemical contaminants of water, *259*
Chemical energy, 26, 92, 94, 99, 199, 200, 201, 202, 203, 205, 206, 227
Chemical evolution, 3, 8, 49, 68, 97, 169
Chemosynthesis, 26, **26**
Chiasma, 155, **157**
Chloride ion, **11**, 32, 50, 115

Chlorination, 261, 262, 265
Chlorine (atom), **10**
Chlorophyll, 41, 50, 93-98
Chlorophyll a, 93, **94**, 96
Chlorophyll b, 93, **94**, 96
Chloroplast, 29, *41*, **42**, 46, 53, 81, 93
Chordate characteristics, *188*
Chordates, *188*, 190, 191, 194
Cilia, 29, *43*, **45**
Cisternae, 38, 39, 42
Citric acid, 102, **102**, 107
Clay, *288*
Clostridium aceticum, 26
Cloth, 283
Clothing, *283*, 284, 285
CO_2 fixation, 94, *97*, **97**
Coacervate, 21, 22
Coagulation-sedimentation, **263**, 264-266
Coal, 200, 201, 228, 229, 231, 232, 270, 275-279
Code unit, of nucleic acids, *69*
Coding system, of nucleic acids, *72*
Codon, *73*, 74, 75, 79-81, **82**, 87, 122, 128, 129
Coenzyme A (CoA) 101, 102, 108
Cofactor, 84, 85, **86**, 96, 101, 103, 104
Coke, 276, 288
Colloid, 31, 33, 261
Colloidal particle, 110
Colloidal suspension, 31, 34
Competitive exclusion, *185*, 186, 193
Complex molecule, *14*, 15, 17, **19**, 20
Composition board, *287*
Compound, *8*, 11, 12-22, 27, 29, 31, 32, 49, 51, 92-94, 98, 106, 107, 114, 199, 202, 203, 206, 207, 217, 219, 223, 227, 243, 244, 247, 249, 250, 257, 260, 261, 263, 264, 276, 277, 286, 287, 296, 299
Concentration gradient, 111-114
Concrete, 231, *286*, 287
Consumer,
  primary, 217, 221
  secondary, 217, 221, 222
  tertiary, 221, 222
Constriction, in cytokinesis, 135, **137**
Controlled (population) curve, 279, **280**, 281

Convergent evolution, *193,* **193,** 194
Conservation, 240
Copper, *288*
Core, of bacteriophage, 298
Corepressor, 89, 90, **90**
Cotton, 283, 284
Covalent bond, 9, *11,* 12, 13, 15, 16, 18, 21, 51, 62, 84
Cover (physical) 283, 286, 289
Cow, 272, 284
Chromatid, 122, **123,** 124, 125, 131, **133,** 135, 136, 139, 141, 142, **143,** 155, 156, 157
Chromatin, 35, 121, 135, 136, **139,** 141, 142
Chromosome, 35, **44,** 45, 53, 81, *121,* 122, **123,** 124–126, 130, 131, **133, 134,** 135, **136, 137,** 139, 140, 141, **141,** 142, **143, 144,** 150, **153,** 155, 156, **156,** 159, 173, 174, **140**
Chromosomal fragments, 123, 125, 126
Chromosome mapping, 156, *157,* 158, 159
Chromosomal mutation, *123,* 124, 125, 171
Chromosomal inversion, 125, **126**
Cristae, 36, **36**
Cropland, 273, 281, 283, 284
Crossover, genetic, *155,* 156, 157, **157,** 158, 173, 174
Crossover, of chromosomes, 142, 155
Crust, of earth, 14
Cyanides, 258, 259, 264
Cybernetics, *208*
Cytochromes, 103, 104
Cytokinesis, 131, *135,* **137,** 139, 142
Cytoplasm, 29, 31, 33, 34, 35, 38, 41, 49, 50, 51, 54, 56, 64, 69, 72, 75, 78, 79, 82, 100, 114, 116, 117, 130, 131, 135, 138, 141, 142, 144
Cytosine, 58, 59, **59,** 63, 69, 70–72, 76, **127,** 128, 129
Dacron, 284, 285, **285**
Dark reactions, of photosynthesis, 94, *97*
Daughter cells, 122, 125, 130, 135, **138**
Daughter chromosomes, 122, **123,** 125, 135, **136,** 139, 141, 142, **143**
Daughter nuclei, 131, **137, 138,** 142
Darwin, Charles, 169
DDD (insecticide), 260, 261
DDT (insecticide), 259, 260, 275

DDT, and natural selection, 179, **180**
Deamination, 108, 127
Death control, 236
Decay organism, 203, 207, 220, 231
Decarboxylase, 105
Deer, 171, 182, 185, 215–220, 223–225
Deficiency (deletion) of genes, 124, 125, 126, 171
  terminal, 124, **124**
  nonterminal, **124, 125**
Degenerate coding system, *73,* 74
Dehydration synthesis, 51, 58, **60**
Deletion of genes, 124, 125
Deletion of single nucleotides, 129
Deoxyribose, 18, 19, **59,** 63, 68, 70, 71, 127
Deoxyribonucleotide, 69
Deoxyribonucleic acid (DNA), 34, 35, 58, 59, **61,** 62, **63,** 67, 68, 69, **70, 71,** 72, 75, 76, 78, 79, 87, 88, 89, 90, 121, 122, 126, 128, 171, 250, 297, 298, 299, 300, 301
DNA, coding system of, *69*
DNA, replication of, *68,* **70, 71,** 75, 76
DNA, stability of, *68*
DNA, synthesis, 75
DNA-polymerase, 69, 71, 75
Depreciation of environment, 252
Desalinization, 265
Detergents, 258, 259
Differential mortality, 182, 183
Differential reproduction, 177, 179, 182
Differential survival, 182, 184, 185
Differentially permeable membrane, 113
Diffusion, *110,* 111, **112,** 114, 116
Diffusion, principles of, *110*
Diffusion through cell membrane, 114
Diffusion through membrane, *112*
Digestion, 107, 117, 218, 219, 220, 227
Dimer, 55
Dinucleotide, 84, 86
Diploid, 123, 144–147
Disaster (population) curve, 279, **280,** 281
Disaccharide, 107
Disease, 219, 223, 224, 232–235
Disease agent, 218, 219, 220, 224
Disorder, measure of entropy, 204–206
Distillation, 265

Index [325]

Divergent evolution, *187*
Division (cell) plate, *135*, **137**
DNA (deoxyribonucleic acid) 34, 35, 58, 59, **61,** 62, **63,** 67, 68, 69, **70,** 71, **72,** 75, 76, 78, 79, 87, 88, 89, 90, 121, 122, 126, 128, 171, 250, 297, 298, 299, 300, 301
Dominant gene, 150, 151, 152, 154, 155, 157, 172, 173, 174, 182
Double helix, 59, 68
Doublet, in coding, 72, 73, 75
Duplication, of chromosomes, *122,* **123**
Duplication, of genes, 124, 125, **125,** 126, 171
Dust, 240
Dustfall, 245, 249, 250
Early earth, 4, 7, 9, 13–16, 92, 169
Ecology, 239, 240, 252, 257, 262, 276, 289, 290
Ecological debt, *252*
Ecological isolation, 181
Ecological niche, 185, 186, 188, 193, 194
Ecosystem, 185, 221, 222, 239, 258, 261, 274, 281, 290
Effects of air contamination, *242*
Effects of noise (on humans), *251*
Effects of smog (on humans), *247*
Effects of smog (on materials), *250*
Effects of smog (on vegetation), *249*
Effluent, 261–264
Egg cell, 123, 144–150, 175
Electrical discharge (early atmosphere), 15
Electrical energy, 199–201
Electricity, 49, 201, 228, 229, 251, 252, 276
Electrodialysis, **263,** 264, 265
Electrodes, **263,** 264
Electrolyte, *49,* 50, 51, 264
Electron, 4, 5, 6, 9, 10, 31, 93–96, 100, 101, 103–105, 303
Electron acceptor, 104
Electron beam, 30
Electron carrier, 84, **86,** 93, 94, **95, 95, 96, 98**
Electron carrier system, 93–96, 103
Electron flow, **96, 98**
Electron shell, 6, 9, 10, 11
Electron subshell, 6, 9
Electron transport (transfer) system, 85, **86,** 95, 96, 98, 100, *103,* **103,** 104, 107

Electronic configuration, 6, 8, 9, 14, 63
Electrons, sharing of, 11, 12
Element, 6, 7, 14, 64
Emphysema and air pollution, 247
Energy flow, 199, 200, **201,** 202, **203,** 204, 206, 220, 221, **232**
Energy Exchange system (of cells), 50
Energy interconversion, 199–203, 205, 206
Energy rich bond, 56, **57**
Energy rich compound, 25, 26
Energy state, 93
Endocytosis, 110, 114, *116,* **116**
Endoplasmic reticulum, 29, 34, *38,* 39, 41
  agranular, 39, **40**
  granular, **38,** 39, **40,** 81
Environment, 27, 28, 64, 68, 92, 93, 97, 110, 128, 147, 169–173, 177–180, 182, 185, 186, 188, 193, 201–203, 207, 214, 215, 218, 223, 226–231, 234, 239, 250–252, 260–262, 265–268, 272, 277, 280, 281, 283, 284, 286, 289, 290, 291, 296
Environmental research, 253
Enzyme, 31, 36, 37, 39, 41, 50, 51, 67, 69, 71, 76, 80, 83–94, 98, 101–106, 206, 296
Enzymatic deficiency, *87*
Enzyme-reactant complex, 83, 84, **85**
Equatorial plane, **134,** 135, **140,** 141, 142, **143**
Equilibrium, 211
Entropy, *204,* 205, 206, 207
Erosion, 231, 281, 284
Ester, 51, 52, 285
Ether, 17, **17,**
Ethyl alcohol, 104, 105, **105,** 257, 258
Ethylene, 244, **244,** 249
Eutrophication, *257,* 258, 261
Evolution, 128, 169, 170, 174, 177, 181, 182, 186, 194, 195, 196
Evaporation, 212, 243, 255, **256,** 264
Experiment, 302
Extracellular fluid, 115
Extracellular (cell) structure, *45*
FAD (flavine adenine dinucleotide) 85, 102, 103, 104, 108
FADH$_2$ (reduced flavine adenine dinucleotide), 102, 103, 104
Farmer-hunter (human) social development, 228, *229,* **230**

Fats, 31, 32, 39, 41, 51, *54*, 55, 64, 92, 106, 107, 108, 110, 117, 256
Fatty acid, 15, 17, *18*, **19**, 54, **54**, 55, 107, 108
Feedback systems, *209*
Fermentation, 100, *104*, 105, **105**
Fiber board, *287*
Fire, 201
First filial generation (F$_1$), 149
Fish, 186, 258, 259, 273, 274
Fission (nuclear) *278*, 279
Flagella, 29, *43*, **45**
Flavine adenine dinucleotide (FAD), 85, 102, 103, 104, 108
Flavine adenine dinucleotide, reduced (FADH$_2$), 102, 103, 104
Flooring, *287*
Floride compounds, 245, 249
Flowering plant, 145, *146*, 147, 196
Food chain, 202, 217, 221–223, 260, 272, 273
Food-fuel, 26, 92, 93, 97, 106, 171, 179, 194, 202, 206, 207, 214, 217, 218, 220, 221, 224, 225, 229, 230, 232, 233–236, 241, 250, 255–258, 264, *270*, 271–275, 279–281, 288, 290, 296
Food production on land, *271*
Food production in the seas, *273*
Foodstuff, 92–94, 99, 107, 257, 274, 275, 301
Fog, 242, 243
Forelimbs, of vertebrates, 186, **186**, 194
Forest, 224, 281, 284, 287, 289, 290
Formaldehyde, 245, 246, **246**, 247
Fossil fuel, 278, 285
Fossil records, 186, 188, 196
Frame-shift mutation, 128, **128**, 129
Fructose, 55, 56, **56**
Fructose diphosphate, 100, **101**
Fructose phosphate, 100
Fuel, 243, 246, 247, 270, *276*, 277, 278, 289
Furanose, 56
Furnace, 210, 211, 242, 243
Functional group, *51*, **52**, 85
Fungi, 203, 220
Gamete, 135, 144, 145, 146, 147, 150, **151**, 152, **152**, **153**, 154, 155, **156**, 157, 173, 174, 175
Galactose, 87
Galactosemia, *87*

Galvanized steel, 288
Gametophyte, 145
Gas (physical state of), 7, 12–14, 111
Gaseous mass, 12, 14
Gasoline, 243, 244, 245, 277
Gasoline engine, 244, 245
Gene, *87*, 88, 122, 123, **123**, 124, 125, 149, 153, 155, 158, 159, 171, 173, 174, 178, 179, 180, 182
Gene flow, 180, **181**
Gene frequency, 175, 176, 179, 182–184
Gene frequency change, *182*, 183, 184
Gene linkage, 155, 156, **156**, 173
Gene pool, 174–182
Generator, of electricity, 200–202
Genetic code, *74*, *75*, 128
Genetic drift, *175*, 176, **176**
Genetic recombination by crossover, *173*
Genetic recombination by meiosis-fusion, *172*, 173
Genetic variability, 128, 169, 180
Genetics, *121*, 148, 174
Genotype, 149, 150, 154, 155, 169, 173–176, 182, 183
Geographic isolation, 180
Gericide, *234*
Ghost, of bacteriophage, 300, **301**
Glass, 231, *287*
Glucose, 18, **19**, 20, 26, 45, 55, 56, **56**, **57**, 58, 97, 98, 100, **101**, 104–108, 114, 205
Glucose phosphate, 100
Glycerine (glycerol), 17, *18*, **19**, 54, **54**, 55, 107, 108
Glycogen, 107
Glycolysis, *100*, 101, **101**, 103–107
Golgi apparatus, 29, *39*, 41, **42**, 81, 117
Gorilla, 191, **191**
Grain crops, 271
Grana, 41, **42**
Grass, 146, 272, 286
Grassland, 224
Green plant, 26, 93, 217
Grit chamber, 261, **262**
Ground cytoplasm, *31*
Guanine, 58, 59, **59**, **63**, 69, 70, 71, 72, 76, **127**, 129
Guanine triphosphate, 103

Index [327]

Half life, *195*, 250, 259
Haploid, 123, 244, 145, 147
Hardy-Weinberg law, 175
Head, of bacteriophage, 297, **298**
Heat, 7, 9, 84, 92, 93, 199, 200-207, 212, 213, 221, 270, 276-278
Helium (atom), 5
Herbivore, 185, 202, 203, 206, 207, 217, 219-224, 229, 230, 232, 233, 272, 273
Herring, 272, 273
Hereditary (genetic) information, 67, 121, 126, 128, 130, 142, 145, 149, 171
Heredity, 148
Heterotroph, 25
Heterozygous (for a gene), 150, 155, **156**, 157
High energy electron, 95
High energy phosphate bond ($\sim PO_4$), **57**, 100
Hooke, Robert, 29
Hominidae (family), 192
Homo (genus), 192
Homo neanderthalensis, 193
Homo sapiens, 193
Homologous chromosomes, *123*, 135, *140*, **140**, 141, **141**, 142, 150, 155, 156, 157, 172
Homologous structure, 194
Homozygous (for a gene) 150-152, 154, 155, 157, 173
Hormone, 116
Host, 219, 300, 302
Human community, 232
Human ecology, 239, 240
Human population dynamics, *226*
Human population explosion, 236, 267, 273
Human society, 227, 228, 229, 267, 270, 291, 303
Hunter (human social development) *228*, 230
Hunting (by man) 224, 225
Hydrogen (atom) 4, **51**
Hydrogen bonding, 62, **62**, 63, **63**, 68, 69, **70**, **71**, **76**, 80, 84, **127**, 128
Hydrogen ion, 50, 51
Hydrogen (molecule) **13**, 15, 276
Hydrogen sulfide, 257, 258
Hydrocarbons, 51, 55, 242-246, 277

Hydrological cycle, 255, **256**
Hydroxyl group, 18
Hydroxyl ion, 50, 51, **52**, 55
Hypothesis, 302
Ion, 11, 31, 32, 49, 50, 110, 111, 112, 114, 115, 264
Ionic bond, *9*, 10, 11
Ionic compound, 31, 49
Inch (in) 30
Inducer, 89, **89**
Inducible operon system, 88, **89**
Individual freedoms (and pollution), 252
Infanticide, 234
Industrial man, 289
Industrial waste, 264, 267, *268*
Industrial waste treatment, 260, *264*, 268, 280
Industrialist (human social development) 228, *229*, **230**
Industrial revolution, 229
Inheritance, 122
Insecticides, 259, 260, 268, 284
Insertion, of genes, 124, 125
Insertion, of single nucleotides, 128, 129
Insulation, 287
Internal combustion engine, 253
Interatomic attraction, 7, 8
Intermolecular attraction, 14, 32
Interphase, meiotic, *139*, **139**
Interphase, mitotic, *131*, **132**, 135
Intracellular fluid, 115
Intrauterine device (IUD), *235*
Inversion, of genes, 125, **126**, 171
Intracellular (cell) structure, *31*
Iron, 231, 270, 286, 287, 288
Isolated (energy) system, 200, 204, 205, *206*
Isolation-speciation, *180*, **181**, 187, 194
Isotope, 7, 298
IUD (intrauterine device), *235*
Kaibab national forest, 224
Ketoglutaric acid, **102**, 103, 107, 108
Kinetic energy, 7, 8, 13, 14, 199-202, 205
Krebs, Sir Hans, 102
Krebs (citric acid) cycle, 100, **101**, 102, **102**, 103, 104, 106-108
Krill, 272, 273

Lack of dominance (genetic), 155
Lactic acid, 106, **106,** 107, 108
Lamprey, 188, **190**
Law of independent assortment (genetic) *152*, 153, 155
Law of segregation (genetic), *149*
Laws of thermodynamics
  first, 199, *200*, 206
  second, 199, *204,* 205
Lead, 245, 289
Leather, *284*, 290
Level of organization, 3
  I, the atom, 4
  II, the simple molecule, 7
  III, the complex molecule, 14
  IV, the macromolecule, 20
  V, the cell, 21
Life, 3, 9, 25, 27, 28, 29, 31, 49, 50, 64, 92, 123, 130, 145, 180, 181, 186, 188, 193, 196, 217, 220, 270, 272, 290, *295*, 296, 297, 302
Life cycle, 130, 144, 145, 147
Life cycle of bacteriophage, *297*, 301, **301**
Light, 9, 26, 93, 94, 96, 97, 98, 99, 199–202, 206, 207, 218, 223, 229, 231, 245, 246, 273, 285
Light reactions, of photosynthesis, 94, *95*, 98
Lightning, 15, 92
Limestone, 286–288
Linen, 283
Liquid, physical state of, 12, 14, 111
Liver fluke, 219
Living, 3, 7, 15, 17, 19, 20, 23, 24, 27, 29, 49, 53, 55, 64, 92, 99, 121, 130, 194, 199, 206, 207, 208, 215, 220, 229, 295, 296, 297
Loci, 122
Lock and key (mechanism of enzymes), 84, **85**
Lumber, *286,* 287
Lumber consumption, 286
Lung cancer, and air pollution, 247
Lysosome, 29, *36,* 37, **37,** 39, 41, 81
Mackerel, 272, 273, 274
Macromolecule, *20,* 21–25, 30, 58, 117
Magnesium, 50, 93, 115
Malaria, 233
Malnutrition, 87, 270, 271

Maltose, 55
Mammal, 186, 190, 191, 194, 196, 268
Man, 144, 171, 188, 190, 193, 196, 202, 226, 227, 230, 232, 239, 240, 243, 255, 257, 258, 270, 283, 285, 296, 297
Map distance (genetic), *158*
Map units (genetic) *158*
Mass transportation, 252
Maternal chromosome, 123, **140,** 141, **141,** 142, 157
Mechanical (energy) system, 199, 200, 208
Megagametophyte, 145, 146
Megaspore, 145, 146
Megaspore mother cell, 145, 146
Meiosis, 130, 138, *139*, 140, 142, 144–146, 148, 150–152, 155, 157, 172, 173
Meiotic cell division, *135*, 139
Meiotic division, first, *140,* 141, 142
Meiotic division, second, *142*
Melanine, 87
Membrane, 29
Mendel, Gregor, 148–150, 155
Mendelian genetics, *148*
messenger-Ribonucleic Acid (mRNA), 79, 81, **82,** *83,* **89,** 90
mRNA-ribosome complex, 79, 80, 81, **82,** **89**
Mercaptan, **52**
Mercury, 258, 259, 260, 261
Metabolism, 212, 213, 218, 219, 220, 227, 270, *295,* 296
Metaphase, of mitosis, *131,* **134,** 142
Metaphase I, of meiosis, *140,* **140,** 141, 142, 153, **153,** 154
Metaphase II, of meiosis, *142,* **142**
Methane, 12, **13,** 15, **16,** 257, 276, 277
Microbiological production of food, *276*
Microgametophyte, 145, 146
Micron (μ), 30, 35, 76
Microorganism, 276
Microscope, 29, 30, 35, 121, 130, 145, 146, 155
  electron, 30, 31, 34, 36, 39, 122
  light, 29, 30, 41, 122
Microspore mother cell, 145, 146
Millimeter (mm), 30, 58, 76
Millimicron (mμ), 30, 31, 37, 41, 174

Index [329]

Minerals, 270, 271
Mitochondria, 29, *35,* 36, **36,** 53, 81, 100–103
Mitosis, 130, *131,* 139, 140, 145, 147
Mitotic cell division, *130,* 139, 144, 146, 147
Modes of living (human), 228, 229, **230**
Mohair, 284
Molecule, 8, 9, 12, 13–15, 17, 21–23, 31–33, 53, 54, 57, 59–61, 63, 64, 71, 79, 80, 82, 83, 104, 105, 110–113, 205, 206, 244
Molecule, complex, *14,* 15, **19**
Molecule, simple, *7,* 14, 15, 21
Molecules, formation of, *8*
Monomer, 20, 23, 24, 285
Monosaccharide, 107
Mountain lion, 219, **219,** 220, 230
mRNA (messenger-ribonucleic acid), 79, 81, **82, 83, 89, 90**
Municipal waste, *267*
Muscle cell, 36, **47,** 48
Mutagen, 128
Mutations, 171, 172, 175, 178, 179, 180, 182, 188, 250
NAD, (nicotinamide adenine dinucleotide), 85, 86, 100–106, 108
NADH$_2$, (nicotinamide adenine dinucleotide, reduced) 100–106
NADP (nicotinamide adenine dinucleotide phosphate) 85, 86, 94, 96, 98
NADPH$_2$ (nicotinamide adenine dinucleotide phosphate, reduced 94, 96–98
Natural gas, 270, 277
Natural resources, 283, 286, 289
Natural selection, 169, 170, 175, 177–180, 182, 185, 194, 296, 297
Neanderthal man, 168, 192, **192**
Negative feedback, 208, *209,* **210, 211,** 212, 213, 214, **214,** 223, 224
Nerve cell, **47,** 48
Net movement, by diffusion, 110–112
Neutron, 4, 5, 7, 278
Nicotinamide adenine dinucleotide (NAD), 85, 86, 100–106, 108
Nicotinamide adenine dinucleotide, reduced (NADH$_2$), 100–106
Nicotinamide adenine dinucleotide phosphate (NADP), 85, 86, 94, 96, 98

Nicotinamide adenine dinucleotide phosphate, reduced (NADPH$_2$), 94, 96–98
Nitrates, 257–259, 263
Nitric acid, 245
Nitrogen (molecule), 8, 240
Nitrogen oxides, 242, *244,* 245, 246, 249
Nitrogenous base, 17, *18,* **19,** 21, 60, 63, 68, 69
Nitrous acid (mutagen) 127, 128
Noise pollution, *251,* 252
Nonpolar compound, 32
Nonrenewable resource, 275, 276, 277
Nonsense codon, 74
Nuclear envelope, 33, 34, **34,** 35, 81, 130, 131, **132, 133,** 135, **137, 139,** 140, 141, 142
Nucleus, of atom, 5, 6, 11, 63, 195, 250
Nucleus, of cell, 29, *33,* 34, **34,** 35, 41, 69, 75, 78, 79, 121, 122, 131, **132,** 135, **137,** 138, 139, **140,** 142, **144**
Nuclear (atomic) energy, 228, 231, 232, 250, 251, 278, 279, 280
Nuclear fuel, 250, 251, 270, 276, 278, 279
Nuclear radiation, *250*
Nucleic acid, 20, **20,** 21, **21,** 23, 25, 31, 34, 35, *58,* 59, **60,** 64, 69, 72, 73, 75, 87, 117, 121, 128, 296
Nucleolus, 34, **34,** 35, 42, 81, 131, **132,** 135, **139,** 140
Nucleoplasm, 34
Nucleoprotein, *121,* 122
Nucleotide, 17, *18,* 19, **19,** 20, 21, 58, 59, **60,** 63, 68, 69, 72, 75, 78, 79, 80, 82, 85, 126, 128, 129
Nutrients, 257, 258, 262, 264, 270, 273, 275, 276
Nylon, 277, 284, 285, **285**
Observation, 302, 303
Ocean, 14, 15, 17, 20, 21, 22, 23, 49, 50, 181, 273, 274, 275
Octane, 243, **244**
Offspring, 67, 68, 148, 149, 152, 153, 157, 158, 170–174, 176, 182, 183, 215, 224, 250
Oil (petroleum), 228, 229, 231, 232, 258, 259, 270, 275, 277, 278, 279, 284, 285
Oil shale, 277, 278
Oil sand, 277, 278

Olefin, 243, 244, **244**
Omnivore, 227
Open seas, jurisdiction, 274
Open energy system, 200, *206*, *207*
Operator gene, *87*, **88,** 89, **89, 90,** 91
Operon, 87, **88,** 90, 122
Order, measure of entropy, 205, 206
Organelles, 29, 30, 31, 33, 36, 37, 41, 43, 46, 93
Organelles of cytoplasm, *33*
Organic acid, 51, **52,** 55, 285
Organic base, 51, **52,** 56, 58, 285
Organic compound, 15, 21, 22, 25, 26, 27, 29, 32, 49, 50, 51, 53, 93, 97, 99, 100, 106, 217, 241, 256, 261, 262, 263, 276, 289, 296
Organic electrolyte, 51, **52**
Organic solvent, 55
Organism, 15, 22-27, 29, 30, 46, 53, 59, 64, 67, 68, 75, 87, 90, 91, 97, 99, 115, 121-125, 130, 145, 147, 149, 150, 155, 169-174, 177, 180, 184, 187-189, 193-195, 199, 200, 202, 206-208, 211-223, 226, 231, 239, 241, 242, 256-258, 260, 262, 274, 289, 296, 297
Origin of life, *3*, 29, 169, 196
Orlon, 284, 285, **285**
Osmosis, 113, **113,** 114
Osmotic pressure, 113
Ovaries, 144, 145
Overlapping coding system, *74*, 75, **75**
Oxaloacetic acid, 102, **102,** 103, 107, 108
Oxygen (atom), 12
Oxygen (molecule) (O$_2$), 8, **12,** 18, 26, 27, 93, 96-106, 114, 205, 206, 240-246, 256-258, 276, 277, 287
Oxygen revolution, *27, 97*
Ozone, 15, 27, 245, 246, 249, 250
PAN (peroxyacetyl nitrate), 245, 246, 247, 249
PGA (Phosphoglyceric acid), 97, **97,** 98, 100, **101,** 107, 108
PGAld (Phosphoglyceraldehyde), 97, **97,** 98, 100, 108
Paraffin, 243, **244**
Parasite, 218-220, 224, 232-236
Parent, 67, 68, 123, 130, 147, 151, 152, 157, 172-175

Paternal chromosome, 123, **140,** 141, **141,** 142, 157
Particulate factor (gene), 149, 150
Particulate matter, in air, 241, 242, 245, 247
Particles, subatomic, 4, 5, 6, 7
Patterns of evolution, *185*
Pea plants, 148-154
Peptide bond (linkage), 53, 54
Per capita food production, 271
Permeable membrane, 113, 114
Peroxyacetyl nitrate (PAN), 245, 246, 247, 249
Persistant contaminants, of water, *259*, 260
Petroleum fuel, 276, *277*
Petroleum products, *289*
Pesticides, 128, 260, 268
Phagocytosis, 116, *117*, **117**
Phenol, 258, 259
Phenotype, 150, 151, 153-155, 157, 158, 169, 173, 179
Phenylalanine hydroxylase, 85
Phenylketonuria, *87*
Phosphate, 18, **19,** 21, 50, **52,** 56, **57,** 63, 64, 68-71, 76, 86, 95, 96, 98, 99, 100, **101,** 103, 127, 257-259, 263, 264
Phosphoglyceric acid (PGA), 97, **97,** 98, 100, **101,** 107, 108
Phosphoglyceraldehyde (PGAld), 97, **97,** 98, 100, 108
Phosphorolation, 100
Phosphorous-32 (radioactive), *299*
Phosphoryl, **52**
Photochemical reaction, 245, 246, **246,** 247
Photochemical smog, *245*
Photolysis, of water, 96
Photophosphorylation
 cyclic, *95*, **95**
 noncyclic *96*, **96**
Photosynthesis, 26, **26,** 41, 55, 85, *91*, 92-94, 96, 97, **98,** 99, 202, 203, 217, 218, 227, 231, 241, 256, 258, 272, 273, 275, 276, 296
Photosynthesis, process of, *94*
Photosynthetic requirement, 93
Phylogenetic tree, 188, **189**
Phylogeny, *188*
Physical environment, 239
Physical state, 8, *12*, 13, 14

Phytoplankton, 272–274
Pigment, *93*
Pinocytosis, *116*, **116**
Pithecanthropus erectus, 192, **192**
Plankton, 274
Plant, 26, 41, 45, 46, 93, 106, 131, 140, 145, 147, 202, 203, 206, 207, 217, 218, 220, 223, 227–233, 246, 247, 249, 256–258, 261, 272, 273, 276, 277, 296
Plant community, 223, 224, 230
Plant products, for clothing, *283*
Plasterboard, *287*
Plastics, 289
Plutonium-239, 250, 251, 278
Plywood, 286, 287
Point mutations, *126*, 128, 171
Polarity, 31, 32, **62**, 63, **63**
Poles, of a cell, 131, **133, 134,** 135, **136,** 140, 141, 142, 150
Pollen grain, 147
Pollen tube, 147
Pollution, 240, 241, 247, 248, 251, 253, 267, 274, 275, 277, 281, 289, 291
Pollution suppression, 252
Polyester, 285
Polymer, *20*, **20,** 21, 23, 24, 45, 54, 55, 58, 67, 284, 285
Polypeptide chain, 84
Polysaccharides, 117
Polysome, 81, 83, **83**
Population increase, potential
    housefly, *170*
    rabbit, *214*
    deer, *216,* 217
    man, *226, 227*
Porcelain, 289
Porpoise, 193, **193**
Positive feedback, 208, 209, *210,* **211,** 212, 213
Potassium (ion), 50, 115
Potential energy, 199, 202, 206
Predator, 179, 182, 214, 218, 219, 224, 225, 230, 232, 233, 280
Precipitation, 255, **256**
Prey, 185, 214, 219, 220
Primary (sewage) treatment, *261,* **262,** 265, 267

Primate, 191, 192
Primitive life form, 25–27, 92, 169, 188
Protein, 20, **20,** 21, 23, 25, 31, 34, 35, 39, 41, **51,** 53, **53,** 54, 58, 64, 67, 69, 72, 73, 75, 78, 80–84, 87–92, 106, 108, 110, 116, 117, 121, 122, 125, 218, 257, 264, 270–273, 276, 284, 297, 299
Protein, importance to cell, *81*
Protein sheath, bacteriophage, 298, 299, 300
Protein synthesis, 35, 39, 41, 50, 64, 69, 72, 74, 75, *78,* 80, **82,** 83, 87, 89, 90, 91, 122, 126, 131
Protein synthesis, control of, *87*
Protein synthesis, mechanism of, *78,* 87
Producer, of food, 217, 221, 222
Production cost, of goods, 252
Proton, 4, 5, 6, 94, 96, 98, 100, 101, 103, 105, 114, 115
Prophase, of mitosis, *131,* **133,** 140
Prophase I, of meiosis, *140,* **140,** 155
Prophase II, of meiosis, *142,* **143**
Punnett square, 150, **151, 152, 154, 173, 175**
Purine, 18, **19,** 58, 59, **59,** 62, 68
Pyranose, 56
Pyrimidine, 18, **19,** 58, 59, **59,** 62, 68
Pyruvic acid, 100, 101, **101,** 102, **102,** 103, 104, 105, 106, **106,** 107, 108
Quinone, 103
Quinone-H, 103
Rabbit, 171, **190,** 191, 213, 214
Radiation, 124, 125, 128
    alpha, 195, 250, 298
    atomic, 15, 124, 128
    background, 250, 251
    beta, 195, 250, 298
    cosmic, 27
    gamma, 195, 250, 298
    X-ray, 15, 124, 128, 195, 198
Radioactive atoms, *195,* 250, 298, 299
Radioactive dating, *195, 196*
Radioactive contamination, 251
Radioactive isotope, 298, 299
Radioactive tracer, *298*
Radium-226, 250
Rain, 14, 255
Random motion, 33, 110, 111, **111,** 112, 205
Range, 216, 224, 284, 290

Rayon, 283, 284
Reactive center (of chemical), 20
Reattachment, of chromosomes, **125, 126**
Recessive gene, 150, 151, 152, 155, 157, 173, 174, 182, 184
Recombination, of genes, 158
Recycling of metals, 288, *290*
Reduction, *94*
Regulator gene, *87,* **88, 89,** 90
Renewable resource, 283, 284, 290
Renewable resources management, *290*
Replication, of chromosomes, 121, *122,* **123, 125,** 130, 139, 142
Replication, of DNA, *68,* **70, 71,** 122, 126, 128, 129, 131, 142
Repressible operon system, *88,* 89, **90**
Repressor substance, 88, 89, **89,** 90, **90,** 91
Reproduction, 67, 123, 130, 145, 169, 172, 184, 214, 215, 217, 220, 296–298, 301, 302
Reproductive cycle
  animal, **144**
  bacteriophage, **301**
  flowering plant, **146**
Reproductive potential, of organisms or populations, 169, *170,* 213, 215, 220, 226, 232, 234, 235
Reptile, 186, 196
Respiration, 85, *99,* 100, 103–108, 115, 241, 256, 296, 297
Responsiveness, 297, 301
Reverse endocytosis, 117
Reverse osmosis, **263,** 264, 265
Ribonucleic acid (RNA), 34, 35, 41, 58, 69, 72, 75, 78–80, 121
RNA-polymerase, 75, 76
RNA synthesis, 69, 75, *76,* **76,** 77
Ribonucleotide, 69, 76
Ribose, 18, 19, **19,** 56, **57,** 59, 76
Ribosome, 29, 34, 35, 38, 39, 40, 41, *41,* 50, 78, 79, 81, **82,** 83
Ribosome-mRNA complex, 81
Ribosome-RNA (rRNA), 78
Ribulose diphosphate, 97
Rigid control, of systems, 208, *209*
Ritual killing, 234
RNA (ribonucleic acid), 34, 35, 41, 58, 69, 72, 75, 78–80, 121

RNA-polymerase, 75, 76
RNA synthesis, 69, 75, *76,* **76,** 77
rRNA, 78
Rubber, 289
Runaway feedback, *211,* 213, 215, 224, 232, 235
Salt, 49, 50
Salts (water contaminants), 258, 259, 263, 265
Schleiden and Schwann, 29
Science, *302,* 303
Scientific method, *302,* 303
Screening, of sewage, 261, **262**
Scrounge (human social development) 228, **230**
Second filial generation (F_2), 149
Secondary (sewage) treatment, 262, **262,** 265, 267
Sedimentation tank, 261, **262,** 264
Seed, 147, 153
Semipermeable membrane, 113, 263, 264
Sensor, of feedback system, 209, 212, 213
Sewage, 257, 258, 261, 262, 267
Sewage treatment, *261,* 262, 263, 264, 267
Sex cell, 123, 135, 245
Sexual reproduction, 144–147, 153, 154, 172, 173
Shark, 193, **193**
Sheep, 284
Shelter, 283, 286, 289
Silica dioxide (in glass), 287
Skin, 212, 213
Sludge, 261–263
Smog, *242,* 243–249, 253
Smoke, 240, 242
Social costs, of pollution, 252
Social (energy) needs of humans, 226, 231, **232,** 233
Social man, 232, 276
Social organization, of man, 226
Sodium (atom), **10**
Sodium (ion), **11,** 32, 50, 115
Sodium chloride, 11, 258
Soil, 203, 230, 231, 273, 284, 285, 289
Solder, 289
Solid (physical state of), 12, 14, 111, 112
Solute, 113, 114

Index [333]

Solution, 31, 32, 50, 113, 116, 243, 245, 264
Solvent, 113, 114
Speciation, 180, 181, 182, 297, 301
Species, 67, 68, 145, 170, 177-182, 186-189, 215, 219-221, 223, 224, 229, 260, 280, 296, 297
Spindle, 131, **134,** 142, **143**
Spindle fiber, **44,** 45, 81, 131, **133, 134, 136,** 140, **140**
Sperm cell, 123, 144-146, 149, 150, 175
Spontaneous mutation, 126, 127, *128,* 172, 180, 181
Spore mother cell, 145
Sporophyte, 145-147
Stainless steel, 288
Starch 20, **20,** 55, **57,** 84, 92, 107
Steel, 276, *287,* 288
Steel production (U. S.), 288
Stigma, of flower, 147
Strontium-90, 250, 251
Structural gene, *87,* **88, 89,** 90
Subscript, *5,* 8
Substrate, 88, **89**
Sucrose, 55, 107
Sugar, *17, 18,* **19,** 20, 21, 55, 56, 58, 68, 69, 92, 93, 97, 107, 205, 270
Sugar-phosphate backbone (of nucleic acids), 68, 71, 75, 76
Sulfate (ion), 249, 257
Sulfhydryl, **52**
Sulfur-35 (radioactive), 299
Sulfur oxides, 242, 243, 245-247, 249, 277
Sulfuric acid, 246, 284
Sulfurous acid, 243
Sun, 26, 27, 93, 202, 207, 218, 232
Superscript, *5*
Suspension, 31, 33
Synthetic (clothing) fiber, 283, *284*
Synthesis of food, *275*
Systems control, *208*
Systems control, of biological systems, *212*
Systems control, of biological community, *223*
Systems control, of mechanical system, *208*
Systems control, of populations, *213*
Systems control, of human populations, 232

Sweat, 212, 213
Table salt, 11, 32
Tail fiber, of bacteriophage, 298
Tail plate, of bacteriophage, 298
Tar, 289
Taxonomic classification, 188, *189*
Technological (population) curve, 279, 280, **280**
Technological society, 255, 276, 278
Temperature, 204, 210-212, 223, 244
Temperature inversion, 248, **248**
Template, 69, 75, 76, 79, 122
Template replication, 68, *75,* 122, 127
Telophase, of mitosis, 131, *135,* **137,** 142
Telophase I, of meiosis, **141,** *142*
Telophase II, of meiosis, *142,* 144
Territorial waters, jurisdiction, 274
Tertiary (sewage) treatment, *263*
Testes, 144, 145
Tetraethyllead, 244, 245
Thermodynamics, 199
Thermostat, 209, **210**
Thymine, 58, 59, **59, 63,** 69, 70, 71, 72, 76, 127
Time pattern of evolution, *194*
Timetable of evolution, *196*
Thorium, 278, 279
Transcription, 75, 76, 78
Transcription, of RNA, *75,* 76, **76,** 78, **89,** 90
transferRibonucleic Acid (tRNA), 79, *79,* 80, 80, 81, 83
tRNA-amino acid unit, 80, 81, 82
Translation, of coding system, *75*
Translocation, 125, **126,** 171
Transitional (point) mutation, **127,** 128
Transpiration, 256, **256**
Transportation and smog, 253
Trees, 45, 284
Trialcohol, 55, 108
Triester, 54, 55
Tripeptide, 53
Triphosphoribonucleotide, 75
Triplet, in coding, 73-75
Triphosphodeoxyribonucleotide, 68, **70,** 71, 75
Triphosphonucleotides, 69

Tritium (hydrogen-3), 251
tRNA (transferRibonucleic Acid), 79, **79**, 80, **80**, 81, 83
tRNA-amino acid unit, **80**, 81, **82**
Trophic level, 217, 221, 222
    first, 217
    second, 217
    third, 217
    fourth, 217
Trout, 193, **193**
Uncharged tRNA, 81, **82**
Undernutrition, 270, 271
Unrenewable resources, 290
Uracil, 58, **59**, 72, 76, **127,** 128, 129
Uranium, 195, 196, 278, 279
Vacuole, 33, 37, 116, 117
Vacuolar membrane, 33
Variability of organisms, *171*, 172, 173, 182
Variability, in organisms and populations, *174*
Vertebrate (animal), 145, 186, 190, 191, 194
Vitamins, 270, 271
Virus, 30, 121, 130, 297–300, 302

Water, 9, 11, 12, **12, 13,** 26, 27, 31, 32, **32,** 33, 49, 50, 51, 54, 55, 57, 60, **62,** 64, 93, 94, 96, 98, 99, 103, 108, 114, 200, 205, 206, 217, 218, 240, 243, 245, 255–258, 260–267, 270, 274–278, 288, 289
Water contaminants, *259*
Water contaminants sources, 257, 258, *259*
Water pollution, 256, 261, 264, 267
Water power, 228, 270, 276
Water purification, *265*, 267
Water recycling, 265, *266*
Water requirements of man, *255*
Water treatment, *261*
Whale, 272, 273
White blood cell, 117
Wildlife, 284
Wood, 228, 231, 232, 283, 287, 290
Work (and energy), 199, 204, 206, 228
Wool, 284, 290
Woman, 144
Yeast, 100, 104, 276
Zooplankton, 272, 273
Zygote, 144–147, 149, 150, 152, 154, 172